FOURIER OPTICS:
An Introduction

Second Edition

FOURIER OPTICS:
An Introduction
Second Edition

E. G. STEWARD, D.Sc., F.Inst.P., C.Phys.
Professor of Physics and Molecular Medicine
The City University, London

ELLIS HORWOOD LIMITED
Publishers · Chichester

Halsted Press: a division of
JOHN WILEY & SONS

First published in 1987 by
ELLIS HORWOOD LIMITED
Market Cross House, Cooper Street,
Chichester, West Sussex, PO19 1EB, England
The publisher's colophon is reproduced from James Gillison's drawing of the ancient Market Cross, Chichester.

Distributors:

Australia and New Zealand:
JACARANDA WILEY LIMITED
GPO Box 859, Brisbane, Queensland 4001, Australia

Canada:
JOHN WILEY & SONS CANADA LIMITED
22 Worcester Road, Rexdale, Ontario, Canada

Europe and Africa:
JOHN WILEY & SONS LIMITED
Baffins Lane, Chichester, West Sussex, England

North and South America and the rest of the world:
Halsted Press: a division of
JOHN WILEY & SONS
605 Third Avenue, New York, NY 10158, USA

© 1987 E.G. Steward/Ellis Horwood Limited

British Library Cataloguing in Publication Data
Steward, E.G.
Fourier optics: an introduction — 2nd ed. —
(Ellis Horwood series in physics)
1. Fourier transform optics
I. Title
535′.4   QC381

Library of Congress Card No. 86-27427

ISBN 0–7458–0128–5 (Ellis Horwood Limited — Library Edn.)
ISBN 0–7458–0186–2 (Ellis Horwood Limited — Student Edn.)
ISBN 0–470–20806–6 (Halsted Press)

Phototypeset in Times by Ellis Horwood Limited
Printed in Great Britain by R.J. Acford, Chichester

'I have a feeling that clarity usually comes when one
struggles a little less and reflects a little more.'

Winslow in C. P. Snow, *The Masters*

To
Hilary and Martin
and some special friends and colleagues:
they have all inspired and encouraged me
— perhaps more than they realize.

Table of contents

6　　　　　　　　　　**Table of contents**

Preface

The principal new features of this edition are the inclusion of a chapter on image formation from projections (medical imaging by computed tomography), several new appendices that broaden the scope of the book, and problems (with answers and guidance for the reader).

The opportunity has also been taken to revise and expand the text in places where students have mentioned difficulty, either in understanding the concepts involved (e.g. coherence) or the mathematics. The mathematics required initially is only at the university entrance level but those alarmed by gaps in derivations need worry no longer; most gaps have been filled in.

Since its first appearance I have used the book each year in a combined course-unit for undergraduates in the following: Control, Instrumentation and Systems Engineering; Electrical Engineering; Electronic Engineering; Physics; Physics and Electronics. The obvious interest shown by engineers at least as much as by physicists mirrors the use of the book elsewhere in colleges and universities both in the USA and the UK. It provides evidence, if any were needed, of the awareness of the role of the principles of Fourier optics in increasingly diverse fields.

Chapters 1–5, with revision in numerous places and the addition of a section on digital computer-based processing, continue to provide the basic treatment in the optical context. The new chapter then deals with X-ray computed tomography, emission computed tomography, magnetic resonance imaging, and ultrasonic computed tomography. These examples of the general theory of image reconstruction from projections, grouped together under the heading 'medical imaging', represent one of the major advances in clinical medicine in this decade. The final chapter, essentially unchanged but up-dated, is concerned with the use of Fourier methods in the study of radiation sources, in spectroscopy and astronomy, by interferometry.

Five appendices have been added. One is on the 'maximum entropy' method of imaging, a relatively new and amazingly powerful, though expensive, technique that has been applied successfully in recent years to optical image reconstruction, medical imaging, astronomy and X-ray crystallography. Students should know of this development, and the appendix explaining the basic ideas in simple terms has been written by Dr S. F. Gull (Mullard Radio Astronomy Observatory, Cavendish Laboratory, Cambridge), a leading authority on this method. Similarly, my colleague Dr J. D. Weaver has contributed an appendix (and Problem 4.5) on the use, in Communications, of the concepts of spectra, convolution and correlation developed in Chapter 4. And my colleague Janet Wolf (Applied Vision Research Laboratory) has written about her most interesting work on vision, in which she shows how a visual effect seen when a pattern of stripes is rotated bears a Fourier transform relationship to the temporal impulse response of the eye in a manner that is analogous to the transform relationship between a Fraunhofer diffraction pattern and the aperture function.

Summary notes on Chapters 1–5 comprise another appendix: a retrospective overview of the main points has proved to be quite useful. Other appendices that go somewhat beyond the scope of the book, but deal with questions often asked by enthusiastic students, deal with the derivation of the Airy pattern, the Fourier transform of a linear array of N δ-functions (only $N = 2$ is considered in the main text), and the spectrum of a truncated wavetrain.

Comments from most lecturers suggest that the general level of the book is about right. A few consider it too elementary but they are counterbalanced by those who take the opposite view!

As before, there is no intention of making this a general textbook on optics; several excellent ones exist already. A detailed account of holography, the Electromagnetic Theory of Light, Kirchhoff's formulation of Huygens' Principle, Fresnel diffraction, polarization, geometrical optics, etc. have deliberately not been included.

It is a real pleasure to record my thanks to those who have helped me in preparing this new edition. (I much regret that I previously omitted to mention helpful discussion with a past colleague, Dr A. W. Gillies, at the time of the early drafting of the first edition.)

I have already mentioned contributions by two experts in associated, specialized fields. In addition, Dr I. M. Mackenzie (Physics Department, University of Manchester, UK) has given me much help in writing §6.3.2 on magnetic resonance imaging, as has my colleague Dr J. P. Weight in writing §6.3.3 on ultrasonic computed tomography, and Dr F. H. Preston (Scientific Research and Development Branch, Home Office, UK) has contributed section (iii) in §5.5.2, on digital computer-based processing. As well as writing the appendix on maximum entropy data processing, Dr S. F. Gull has expertly guided me in revising the section (§7.6) on astronomy and has provided Fig. 7.18. My colleague Professor V. E. Price (Department of Computer Science, The City University) has kindly produced computer drawings for Figs. 4.07 and G.01, and Professor I. Isherwood (Department of Diagnostic Radiology, Medical School, University of Manchester, UK) the illustrations (Figs 6.06 and 6.08) on medical imaging. I am indebted again to Dr R. S. Longhurst, this time for allowing me to use him as a sounding board for improvements in the main body of the text. Finally, my thanks to Professor Eugene Hecht for his kind remarks in his review of my book and for his helpful suggestions, which I have been able to take into account in preparing this new edition.

London, May 1986   E. G. Steward

Preface to the first edition

This book is intended for students who seek a simple introduction to the Fourier principles of modern optics and an insight into the similar role they play in other branches of science and engineering. Fourier transforms, associated with the operations of convolution and correlation, form the basis not only of image formation and processing with lens systems, but also of studies ranging from the atomic structure of matter to the galactic structure of the universe, and some modern methods of spectroscopy. They also apply to the communication and information sciences of electrical engineering, including the processing of information that is not optical in origin. As a supplement to the undergraduate mainstream textbooks on optics the book bridges the gap to the more advanced treatises in the various specialized fields.

Incoherent imaging is described from the point of view of convolution and transfer functions, with emphasis on the features of linearity and invariance shared with many types of electrical network (non-linear systems being outside the scope of this book). The double Fourier transformation process of coherent image formation is illustrated with particular reference to its application in X-ray crystallography.

As an introduction to the wide range of applications of optical filtering and image processing the basic ideas of amplitude-, phase- and holographic filtering are described, with illustrations drawn from optical and electron microscopy, and from the developing field of pattern recognition. Energy-spectrum correlation and geometrical optics-based processing are also briefly described.

Historical aspects are mentioned wherever they particularly help to convey the spirit of science and show how developments take place: to understand fully the present state of a subject, and where it is going, it is essential to know where it has come from. For example, in Chapter 6[†] it is

---

[†] Chapter 7 in the second edition.

explained how the modern uses of interferometry in radio and optical astronomy, and spectroscopy, have their origins in Michelson's stellar and spectral interferometers: the methods that Michelson devised, and his brilliant awareness of their potential, form the natural starting point for understanding their present-day counterparts.

To encourage and help the reader to consult the whole span of the literature the text is referenced throughout, in addition to the inclusion of a bibliography.

I have assumed that the reader already has a knowledge of mathematics up to the usual university entrance level, all additional mathematics being developed in the context in which it is used. In the main, the treatment is restricted to one dimension so as to keep the mathematical statements as visually digestible as possible.

It is a great pleasure to record my sincere thanks to Dr R. S. Longhurst for stimulating discussions and suggestions at many stages in the preparation of the final manuscript: to be able to tap his experience as author of a major undergraduate textbook on optics has been invaluable. Similarly, I am grateful to Professor Henry Lipson who kindly read and commented on Chapters 1, 2 and 5, and who has been a source of encouragement to me over the years; and to Professor Sir Martin Ryle for helpful comments on the section on astronomy in Chapter 7. My thanks also go to my colleague Professor M. A. Jaswon (Head of Mathematics Department at The City University) for helpful comments on my presentation of some of the mathematics.

I am grateful to Roger Meggs (Physics Department) for assistance in the preparation of the figures, and to Peter Clover (Audiovisual Aids) for making all the ink-drawings.

February 1983   E. G. Steward

J. B. J. Fourier
1768–1830

Jean Baptiste Joseph Fourier was born on 21 March 1768 in Auxerre, about one hundred miles south-east of Paris. By the age of nine he had lost his father (a tailor) and his mother, and was sent to the local military school. There, he became fascinated by mathematics. However, the army was the career open to him, though he was apparently not accepted and went instead to a Benedictine school at St. Benoît-sur-Loire. The Revolution upset his further plans and in 1789 he returned to Auxerre to teach at the school where he had been a pupil.

The political upheaval in France greatly affected Fourier's career. In 1794 he was arrested for courageously defending victims of the short-lived Jacobin 'Reign of Terror'. When released he went to study at the École Normale in Paris and, a year later, became the equivalent of an assistant lecturer at the newly opened École Polytechnique. But in 1798 he was elected to join Napoleon's Egyptian campaign as scientific adviser. After holding various diplomatic posts and virtually governing half of Egypt, though still continuing to pursue his interests in mathematics, he returned to France in 1801, wanting to resume his academic work at the École Polytechnique. Because of his administrative skills, however, Napoleon then appointed him to the 'prefecture' of the Isère *departement* (region), based at Grenoble, and later to the prefecture of the Rhône *departement* centred at Lyons. In 1808 Napoleon conferred a barony on him but by the time of Napoleon's defeat in 1815 he had resigned his title and the Rhône prefecture in protest against the severity of the régime, and had again returned to Paris.

Fourier had done some of his best work at Grenoble, on the mathematical theory of heat conduction. It was in connection with this that he developed the series named after him.

After initial obstruction by Louis XVIII, Fourier was elected to the Académie des Sciences in 1817. A year later he succeeded Laplace as President of the Council of the École Polytechnique.

Fourier made numerous other contributions to mathematics (including the introduction of the $\int_a^b$ notation we use today), some serving as a starting point for the much later work on the theory of functions of a real variable. (From the age of sixteen he had been especially interested in the theory of equations.) But even without these, his mathematical analysis of heat propagation justified his being accepted as the leader of the French analytical school.

Fourier died in Paris on 16 May 1830 from an illness, possibly myxoedema (hypothyroidism) contracted during the Egyptian campaign.

(Based on Gillispie, C. C. (ed.) (1972) *Dictionary of scientific biography*, Scribner, New York, and Williams, T. I. (ed.) (1982) *Dictionary of scientists*, Black, London.)

1

Preliminaries

1.1 INTRODUCTION

This book is primarily concerned with the Fourier aspects of two inter-related topics in physical optics: (i) the formation and processing of images and (ii) the study of the spatial distribution and spectral composition of radiation sources. They are brought together within the same covers because they are inter-related in a number of ways, and when such relationships exist there is, usually, as here, an increased benefit to be obtained from studying the topics together. Their wider implications, referred to in the Preface, will be introduced as they arise.

As we work through these pages we can think of both topics in the context of visible light, but we also deal with some important applications in other parts of the electromagnetic spectrum. In image formation this takes us from the optical region to the way in which X-rays are used to deduce the atomic structure of matter, and at the other extreme, to astronomy and the structure of the universe. In spectroscopy, Fourier-based methods are now used over a wide spectral range.

The mathematical methods named after **J. B. J. Fourier** (1768–1830) are extremely powerful in these topics. They are introduced in Chapters 3 and 4, and are used extensively in the closing chapters. For the mathematical representation of light the '**scalar-wave approximation**' as described in general physics textbooks is entirely adequate and is employed throughout. Appendix A gives a resumé of the notation and basic equations used, together with reminders of the meaning of such terms as path difference and phase difference, and the use of phasor diagrams for summing waves having different amplitudes and phases.

In this chapter we identify, in general terms, some of the physical processes involved in the two main topics. These can be introduced very conveniently in the context of Young's experiment and the phenomena of the Newton's rings type.

The reader will already know that the historical experiment performed by **Thomas Young** in 1801 provided crucial evidence in support of the wave theory of light. We must start by reminding ourselves of some of the details of this experiment and of our present-day interpretation of it.

Figure 1.01(a) shows the arrangement. Sunlight passing through a

Fig. 1.01 — Young's experiment.

pinhole S illuminated an **aperture mask** (or screen) some distance away, containing two closely separated pinholes B and C. On another screen, about as far away again, fringes of light and dark were observed in the region of geometrical shadow around O. Neither pinhole alone produced the fringes and their existence was interpreted as interference between light diffracted at the two pinholes. It will be recalled that according to **Huygens' principle** as developed by **Fresnel** and **Kirchhoff**, every point of an advancing wavefront is regarded as acting as a source of '**secondary wavelets**' whose envelope forms the profile of the wavefront as it advances. When the advancing wavefront is intercepted by a screen containing an aperture, **diffraction** occurs, the wavelets that pass through the aperture having a wavefront envelope that spreads into the region that would be unilluminated shadow according to the **ray theory** of **geometrical optics**. This is indicated in Fig. 1.02(a), which can be thought of as representing one of the apertures in Young's experiment. At any point, P, say, the illumination is the resultant of the **interference** between the wavelets arriving there from all points across the aperture, arriving with different phases due to the different path lengths they have travelled. The pattern on the screen is of the familiar **Fresnel** type described in the standard texts. The details need not concern us at the moment since, if the pinholes in Young's experiment are small enough, the diffracted light from each pinhole, alone, would produce quite a large patch of fairly uniform illumination on the screen. It is the interference

Fig. 1.02 — Diffraction: the spreading of Huygen's wavelets into the geometrical shadow.

between the light diffracted in this way from the two apertures in the experiment that is responsible for the fringes.

Two asides need to be made here. One is to note that the term 'interference' describes what is solely a summation, in accordance with the **principle of superposition**, of displacements where wavetrains overlap; every wavetrain can proceed beyond the region of overlap quite unaffected. In the present context, however, instead of calling the overall pattern on the screen (whatever the number of apertures) an **interference pattern** it is often referred to instead as a **diffraction pattern** in recognition of the physical process by which the light leaves the apertures to arrive at the place where interference occurs. The other point to note is that when the term interference is used in the present context it can be classified as **interference by division of wavefront**. This distinguishes it from **interference by division of amplitude** which occurs in the formation of **Newton's rings**, for example (§1.4).

Before considering the implications of Young's experiment in relation to the subject of this book it will be useful to note some details concerning the fringe pattern itself.

Assume for the moment that monochromatic light, wavelength λ, is used. With the arrangement shown in Fig. 1.01(a), if A_1 and A_2 are the separate amplitudes of illumination at P originating from apertures B and C, then the resultant, R, of their superposition at P can be obtained by using the phasor diagram (Appendix A) shown in Fig. 1.01(b), where δ is the phase difference corresponding to the path difference l and is given by

$$\delta = \frac{2\pi}{\lambda} l .$$

The **intensity of illumination**, or **flux density**†, at P is

$$I = R^2 = A_1^2 + A_2^2 + 2A_1A_2\cos\delta$$
$$= A_1^2 + A_2^2 + 2A_1A_2\cos\frac{2\pi l}{\lambda} \qquad (1.01)$$

For the sake of generality we will not put $A_1 = A_2$, although that would be justified here.

The third term is the 'interference term', and it enables the positions of fringe intensity maxima to be calculated. At points where P is such that $l = 0, \lambda, 2\lambda, \ldots$, the wavefronts diffracted from the two apertures are in phase and we have intensity maxima, with

$$I_{max} = A_1^2 + A_2^2 + 2A_1A_2 = (A_1 + A_2)^2 . \qquad (1.02)$$

Between these, where $l = \lambda/2, 3\lambda/2, 5\lambda/2, \ldots$, the wavefronts are in opposite phase (they are 'out of phase'), and there are intensity minima with

$$I_{min} = (A_1 - A_2)^2 . \qquad (1.03)$$

(For positions of P increasingly away from O the intensity of the pattern is progressively attenuated because the amplitudes of the secondary wavelets diminish with increasing angle. If they did not do so, there would be a wavefront travelling in the backward direction. To deal with this, an **obliquity** (or **inclination**) **factor** — explicit in **Kirchhoff's** analysis — is incorporated in quantitative equations for amplitude/intensity distributions. There would also be a factor allowing for the **inverse square law**, because the distance between P and the apertures varies with the position of P on the screen. In keeping with general practice, such correction terms are omitted from the equations in this book.)

The *spacing* of the fringes, Δ in Fig. 1.01(a), in the central region is given approximately by writing

$$\left.\begin{array}{l} \Delta \approx L\theta_1 \\[1em] D\theta_1 \approx l = \lambda \end{array}\right\} \text{ for first fringe from axis}$$

whence

$$\Delta \approx \frac{L\lambda}{D} \qquad (1.04)$$

It is important to note that the fringe spacing is proportional to λ and inversely proportional to D; also that the actual intensity at any point in the pattern is determined by the amplitude arriving there from each pinhole and the way the two contributions interfere.

There are several respects in which Young's experiment illustrates the types of physical phenomena that are involved in the two main topics of this

---

† Flux (power) received per unit area. Alternative terms: **illuminance**, **irradiance**.

book mentioned at the beginning of this section. Consider the topic concerned with the spectral and spatial distribution of radiation sources. If the fringes in Young's experiment are to be of good 'visibility' — good clarity — it is essential to use a very small source for the illumination of the apertures. The sets of fringes resulting from light leaving spatially separated points in anything but a very small source would be so displaced relative to each other that the nett fringe pattern would be of low visibility.

Matters concerning fringe visibility permeate much of what lies ahead in this book and we need to note how visibility is defined for analytical purposes, without delay. Figure 1.03 shows the type of effect we have just

Fig. 1.03 — The smearing of fringes leads to a fringe pattern of the same repeat but reduced visibility.

described. Displacement of fringes over a small range is depicted in (a), though three distinct fringe patterns are shown whereas in reality there would usually be a continuous spread over whatever is the range (the details would depend on the size and intensity distribution of the source). The nett effect is represented in (b) and the **visibility** is defined as

$$V = \frac{I_{max} - I_{min}}{I_{max} + I_{min}} \left(= \frac{(\alpha + \beta) - (\alpha - \beta)}{(\alpha + \beta) + (\alpha - \beta)} \right) = \frac{\beta}{\alpha} \qquad (1.05)$$

Alternative terms are **modulation** or **contrast**. (Note, however, that 'contrast' is sometimes defined as $(I_{max} - I_{min})/I_{max}$.)

In addition to using a source of small size, the illumination used in Young's experiment needs to be what is loosely called 'monochromatic' if fringes of good visibility are to be observed.

Fringe visibility thus contains information about both the spatial and spectral nature of the source S. It is the extraction of this information from interference effects that will occupy a major part of our thinking and be brought to a head in Chapter 7.

Turning to the other topic, which concerns image formation, radiation is there used as a vehicle for conveying information about an object to the place where the radiation is distributed in such a way as to from an image of the object. With a suitable lens the light leaving the two pinholes in Young's experiment can be used to produce just such an image, of the pinholes. (Similarly, when the pinholes are viewed directly by eye it is the lens in the eye that produces an image on the retina at the back of the eye.) How does the lens function? Why is it that with a large source at S fringes are not seen on the screen yet the insertion of the same lens will still produce an image of them?

The answers to these questions, and the general aspects of the relationships touched on above, between fringe visibility and the nature of the source, involve not only diffraction and interference but also the '*coherence*' of the radiation, to which we must next turn our attention.

1.2 COHERENCE AND LIGHT SOURCES

For interference effects to be observed on the screen in Young's experiment it is necessary that the **wavetrains** of light arriving there from the two apertures B and C in Fig. 1.01(a) should overlap (and of course have the same frequency), and that there should be a constant phase difference between them. If these conditions were ideally met, the illumination at the apertures would be said to be **coherent**.

Conventional light sources are not perfect, however, and they provide illumination that is to a greater or lesser extent coherent — illumination that is **partially coherent**.

Consider the first requirement for coherence mentioned above. From each wavefront incident at the aperture mask containing the pinholes B and C in Young's experiment, new wavefronts spread out from B and C. One visualizes the parts of the wavefronts that travel the paths from B and C to P as **wavetrains**. If the wavetrains are to interfere at P they must not only be of the same frequency but must overlap there, and this will only happen if their length is large compared with the difference in the distance they travel i.e. CP–BP: otherwise they could miss meeting each other at P. We say that the light must have sufficient **coherence length** $l_c = c \cdot \Delta t$, where Δt is called the **coherence time** of the illumination (c is the velocity of light). This is a description of the '**temporal coherence**' of the illumination and it is related to the spectral composition of the light (§1.2.1). Coherence time/length of the illumination used in interference experiments is a key factor in determining fringe visibility.

For the second requirement, that there be a constant phase difference between the wavetrains arriving at B and those arriving at C, the light source must be small enough to ensure that there is a negligible spread, compared with the wavelength of the light, of the distances travelled from different points in the source, to B and C. This is the requirement for '**spatial coherence**' of the illumination field over the width BC (§1.2.2).

The use of a small, common source, S, for the illumination of both pinholes in Young's experiment, goes some way to meeting the two requirements for coherence. They are considered in a little more detail in the following sections.

1.2.1 Temporal coherence

By its very nature, the **photon** (light quantum) emission of light from atoms means that each wavemotion associated with the emission of a photon is radiated for a finite time and the wavetrains therefore have finite length. This means (and we examine it in some detail in §4.6) that it cannot have one single frequency: it would need to be of infinite length for that (§4.6 and Appendix G). Instead, it has a range (albeit narrow) of frequencies, i.e. it has a '**frequency bandwidth**'. Even **laser light** has a finite bandwidth and the ordinary commercial He–Ne laser has a coherence length of a few tens of centimetres. With conventional, non-laser, sources — commonly referred to as **thermal sources** — thermal vibrations of the radiating atoms, together with other effects, degrade the quality of the light and restrict the time during which a wavetrain can be regarded as approximating to a simple harmonic wave. For these reasons the so-called **monochromatic** light from sources such as gas-discharge tubes is more appropriately called **quasi-monochromatic**. **White light** is at the opposite extreme to laser light, with wavetrains so short that no one particular frequency can be identified.

This aspect of the light from any source refers to its temporal coherence, which can be identified qualitatively for the moment as the interval of time during which the phase of the wavemotion changes in a consistently predictable way as it passes a fixed point in space: the longer the time, the greater is the degree of temporal coherence. The length of wavetrain meeting the requirement is the **coherence length**, equal to the product of the coherence time and the velocity of the light as mentioned above. (When the term 'coherence length' is used remember that reference is being made to the spectral purity of the light, not to some aspect of its spatial distribution.)

Even if S in Fig. 1.01(a) were a true 'point source' the effect of finite coherence length means that, for points P progressively further away from the axis, l may assume a value comparable with the length of the wavetrains. Wavetrains that leave B and C simultaneously (and originate from a single wavetrain from S) would not then fully overlap when they arrive at P, and the visibility of the fringes would consequently be reduced. Further along the screen the fringes could completely disappear to give a continuous level of illumination due to the separate, independent contributions from the two apertures.

1.2.2 Spatial coherence

The radiation processes at separate points in any conventional thermal light source are independent of each other, and in that sense such sources may be thought of as incoherent. But our interest is not so much in the nature of the source itself as in the quality of the illumination field it produces, for example, in a plane at some distance from the source. Thus in Young's experiment we are interested in the extent to which there is a constant phase relationship between B and C (Fig. 1.01(a)) so that interference effects can be observed. We have noted the effect of limited temporal coherence, due to the finite bandwidth of the light from a source. What effect does the finite size of the source have?

Figure 1.04 shows schematically and on a grossly exaggerated scale, the two apertures B and C of Young's experiment, with a light source of width W at distance r. Assume that the light has unlimited coherence length. Light from some point S in the source illuminates the apertures, and interference fringes are produced on the screen. According to the location of S there is a certain delay between the time of arrival of any wavefront at aperture B and its arrival at aperture C. The magnitude of this delay determines the positions of the intensity maxima and minima of the fringes on the screen (not shown in this figure) due to the light from S. If the source could consist of just this single point (as in an idealized Young's experiment) then fringes of maximum visibility would be observed. A real source is of finite size, however, and the fringes produced by illumination from other points are displaced relative to those due to S. Furthermore, with a conventional thermal light-source, and no matter how great the temporal coherence, the fringe patterns produced by light leaving different places in the source are purely additive in terms of intensities (i.e. there is no interference between them) since they are quite unrelated. Light from the whole source therefore produces a spread of fringes as illustrated in Fig. 1.03, with a consequent reduction in the visibility of the fringe pattern.

In general, then, the illumination field from a finite source, even if truly monochromatic, is again partially coherent, this time spatially — owing to the spatial distribution of the source. And the degree of coherence between two points in the field is physically manifest in the visibility of interference fringes produced by light from those two points. The actual relationship between spatial coherence and the fringe visibility is dealt with in §7.4.

For practical purposes, one can say that a high degree of spatial coherence is obtained at the aperture screen in Fig. 1.04 if BS–CS only changes by a small fraction of a wavelength for all positions of S from S_1 to S_2. This can be expressed as

$$(BS_2 - BS_1) - (CS_2 - CS_1) \ll \lambda .$$

As W must obviously be very small compared with r, the above approximates to

$$2W\theta \ll \lambda .$$

Since $r\theta \approx D/2$ and $r\phi \approx W$ we then have

$$\phi \ll \frac{\lambda}{D} , \qquad (1.06)$$

which tells us that for the illumination at the pinholes B and C to be spatially coherent, the angle ϕ (in radians) subtended by the source at the pinholes must be appreciably less than λ/D. Conversely, we can say that in the plane of the pinholes the illumination is effectively coherent spatially over a

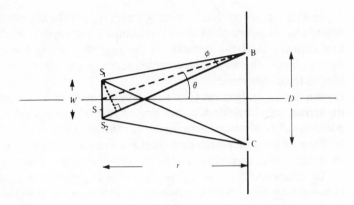

Fig. 1.04 — Spatial coherence. The effect of source size in Young's experiment.

distance D equal to a small fraction of λ/ϕ, the fraction being specified according to any particular experimental requirement. That particular value of D is called the **(transverse) coherence width** of the illumination in the plane concerned. Given sufficient information about the source the **coherence area** in a plane at a specified distance from the source can be defined in the same way.

It is interesting to pause here and realize that no matter how drastically the coherence of an illumination field may differ from one point to another, such differences are not visible to the eye; the eye is only sensitive to intensity.

To return to the reduction in fringe visibility that occurs as source size increases, this provides the basis of **Michelson's stellar interferometer** method for measuring the **angular diameters** of stars (the angles they subtend at the Earth) that are too small to be measurable with a telescope in the conventional way. This is outlined in Chapter 7 where we also see how the actual way in which the visibility changes as the separation between the two apertures is altered reveals information about the brightness distribution of the source.

Finally, we should note that laser light sources are an exception with respect to spatial coherence just as they are with temporal coherence (§1.2.1). There is spatial coherence right across a laser beam (and its width can be increased for practical purposes by means of a suitably designed lens system — a **beam expander** — without loss of coherence). Closely similar to laser light, from the point of view of spatial and temporal coherence, are the **radio waves** emitted by radio transmitters.

1.3   OPTICAL IMAGE FORMATION

We are now in a position to identify and appreciate some of the general aspects of optical image formation, before embarking on their detailed study leading to Chapter 5.

In §1.1 we noted that in Young's experiment the introduction of a lens would enable an image of the two pinholes to be formed. It is evident, then, that diffraction at the pinholes is the first step in the formation of their optical image, the second step being the 'recombination' of that light by the lens to form the image.

We also recalled in §1.1 that fringes are not observed in Young's experiment if a large light source is used. Yet it is a common experience that a lens would still produce an image of the apertures.

Does the lens act differently in these two cases? The short answer is no, for reasons that can fairly readily be appreciated in qualitative terms.

Consider the first case, when Young's experiment is performed in the classical way with what we can describe as reasonably coherent illumination at the aperture screen. The source is small enough to ensure that the same phase relationships exist between every pair of diffracted wavefronts leaving the apertures B and C. Every pair contributes to the same set of fringes, and the intensity maxima and minima in the fringe pattern can be located by reference to Eqn (1.01).

In the second case, using a large source, there is no spatial coherence in the illumination at the aperture screen (we can assume that the temporal coherence is adequate in both cases). As we have seen, fringes resulting from wavefronts leaving one point in the source are displaced with respect to fringes produced by wavefronts originating from other points in the source. If the source is large enough the nett result is a relatively uniform patch of illumination on the screen. Even so, fringes are indeed formed as in the first case, but they are instantaneous and their positions continually change, giving uniform illumination when viewed over even the shortest practicable time. That the information is there, with individual phase relationships still existing in the light that will enter the lens when this is inserted, is evident from the fact that it is still possible to form an image of the aperture(s) with the lens. (When coherent illumination is used the stability of the phase relationships makes it possible under some circumstances to record the full information, including phases, in the wavefronts leaving the aperture screen; this is the basis of holography (§5.4).)

The above comments concerning the imaging under coherent illumination (**'coherent imaging'** for short), of an object in the form of the pinhole mask in Young's experiment, apply equally to (i) more complicated object masks such as the 35-mm transparencies used in slide projectors, (ii) opaque objects illuminated with conventional thermal light, and (iii) 'self-luminous' objects which are luminescent (e.g. a TV picture) or incandescent (e.g. infra-red photography of hot objects). The same *instantaneous* phase relationships exist in each of these categories, just as we have described above.

In this wider context of imaging the term 'scattering' is often used instead of 'diffraction'. There is some confusion in making a distinction between the two in the literature but the details do not affect our deliberations and we can regard the terms as synonymous.

There are two aspects of image formation that now need to be noted.

One is concerned with the formation of a Fraunhofer-type diffraction pattern in the back focal plane of the imaging lens, the other with the effect that the finite aperture of the lens has on the image. They lead to two useful models of image formation (Fig. 1.08) that we explore in Chapter 5.

1.3.1 Fraunhofer diffraction

Consider Fig. 1.05 which depicts an **object mask** containing two very small

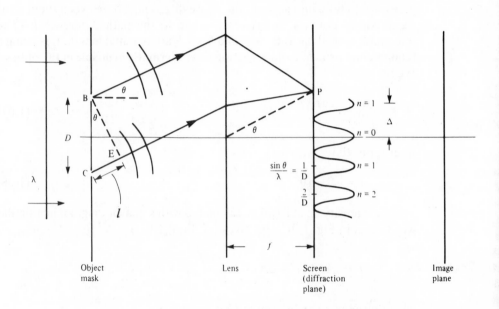

Fig. 1.05 — Diffraction as an intermediate step in image formation.

pinhole apertures, B and C, uniformly illuminated with quasimonochromatic light from a distant source. Plane waves arrive at normal incidence to the mask, and spherical wavefronts spread from B and C. This is the same as in Young's experiment except that in addition we now have a lens producing an image of the pinholes in a plane located as shown in the figure. It is the back focal plane of the lens, however, that is of immediate interest. Consider any point P, at θ to the lens axis; only contributions travelling in direction θ from B and C are brought together there, to interfere (cf. Young's experiment where the interference at P in Fig. 1.01 is between light travelling in *different* directions from the apertures). We shall see that the particular diffraction pattern (defined below as of the Fraunhofer type), formed in the back focal plane of the imaging lens, is a particularly important intermediate step in the formation of the image produced by the lens. It provides valuable insight into the final stage of image formation, and an opportunity of special and unique significance for modifying the image. These are among the topics of Chapter 5, but here we will examine some of the features of the pattern formed in the example we have just described. First, however, note that for such Fraunhofer diffraction patterns to be experimentally accessible it is necessary to arrange that the static phase relationships provided by coherent illumination exist — see the comments in the previous section about the difference between coherent and incoherent imaging. Until Chapter 5, where those differences are again discussed, we shall — unless otherwise stated — make the assumption that coherent conditions exist.

To calculate the resultant of the two contributions at P in Fig. 1.05, it is necessary to take into account the phase difference, δ, between them — as we did with Young's experiment — due to the path difference l. (The remaining optical paths $B \rightarrow P$ and $E \rightarrow P$ are of equal length, this being a fundamental principle of lens action — cf. **Fermat's principle** in ray optics.) We have

$$\delta = \frac{2\pi l}{\lambda} \tag{1.07}$$

where, this time,

$$l = D \sin\theta . \tag{1.08}$$

The intensity, I, at P can be calculated with a phasor diagram in a similar way to that in Fig. 1.01. This gives (cf. Eqn (1.01))

$$I = A_1^2 + A_2^2 + 2A_1 A_2 \cos\frac{2\pi l}{\lambda}$$

where l is now given by Eqn (1.08).

The two amplitudes are equal since the same direction, θ, is involved for both contributions. Putting $A_1 = A_2 = A$ the above equation becomes

$$I = 2A^2\left[1 + \cos\left(\frac{2\pi}{\lambda}D\sin\theta\right)\right]$$

$$= 4A^2\cos^2\left(\frac{\pi}{\lambda}D\sin\theta\right) . \tag{1.09}$$

The pattern is commonly referred to as consisting of '**cos² fringes**'.

In the direction of a fringe intensity maximum the illumination from each aperture is in step and

$$l(= D\sin\theta) = n\lambda \tag{1.10}$$

where n, the **order of diffraction** (or interference), is an integer or zero. The intensity is then

$$I_{max} = 4A^2 .$$

It should again be noted that although the directions of the maxima are determined by D and λ, their magnitudes are controlled by the value of the amplitude transmitted in those directions by the individual apertures (dealt with in §2.3).

Between the maxima there are minima, which are zero only if (as is usual) $A_1 = A_2$.

The fringe spacing, Δ, is readily calculated with the aid of Fig. 1.05. If the first-order maxima are at $\theta = \pm \theta_1$ say, and if θ_1 is small (the figure is of course grossly exaggerated) we have

$$\Delta \approx f\theta_1$$

$$D\theta_1 \approx \lambda$$

whence

$$\Delta \approx \frac{f\lambda}{D} \, . \tag{1.11}$$

Regardless of the shape, number, etc., of apertures in the mask, the patterns formed in this way are all described as being of the **Fraunhofer** type. The arrangement is shown again in Fig. 1.06, with an object mask consisting

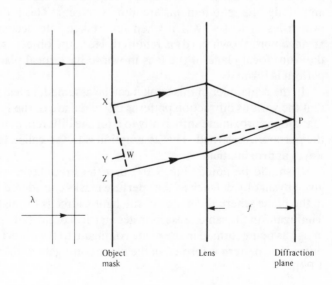

Fig. 1.06 — Fraunhofer diffraction.

of a single aperture of finite size; the details of the Fraunhofer pattern this gives are dealt with in §2.2. In all examples of Fraunhofer diffraction there is a linear variation of the optical path length travelled by diffracted light from points across the object screen to any particular point in the pattern. Thus in Fig. 1.06 the optical path difference YP–XP = YW is proportional to XY. In contrast, the corresponding variation in Fig. 1.02 is non-linear, and patterns formed under those conditions are of the **Fresnel** type.

Since with a Fraunhofer diffraction pattern the same pattern would be obtained at infinity if the lens were removed, an alternative description often used is **far-field diffraction**. In contrast, Fresnel diffraction is then referred to as **near-field diffraction**, though it must be stressed that a wide variety of pattern falls into the Fresnel (near-field) category, with the Fraunhofer type at one extreme. For example, when Young's experiment is performed with both the source and the screen (on which the fringes are observed) at appreciable distances from the aperture mask the pattern is practically indistinguishable from the Fraunhofer type. If the distances are substantially reduced (as suggested in the exaggerated scale of Fig. 1.01(a)) the Fresnel pattern would be quite different.

It is not necessary here to go into the vast subject of the methods of illumination used in optical microscopy. But it is useful to note in Fig. 1.07 some of the arrangements used specifically for observing and recording Fraunhofer diffraction patterns — often for 'optical filtering' purposes (Chapter 5). Of these, (a) is another way of providing the same illumination as in Figs. 1.05 and 1.06, whilst (b) and (c) produce almost the same patterns as (a). When accurate details are required concerning both phase and amplitude, the requirements are more stringent. This is due to the imaging properties of lenses, which when fully taken into account show that the arrangement shown in (d) is required. Here the object screen is located in the front focal plane of the lens in whose back focal plane the diffraction pattern is formed.

In the remainder of this book it can be assumed, unless otherwise stated, that the various diffraction patterns discussed are of the Fraunhofer type or that they approximate sufficiently to it for the differences to be negligible for the purpose concerned. This assumption is the so-called **far-field** (or **plane-wave**) **approximation**.

It should be noted (Fig. 1.07) that the Fraunhofer diffraction pattern given by an object (such as the aperture masks considered here) is observed in the plane where the image of the light source is formed. Using the term '**conjugate**' in the same sense as in geometrical optics (where one refers to an image as being formed in the plane conjugate to the object), the Fraunhofer diffraction pattern is formed in the plane conjugate to the light source (Fig. 1.08).

1.3.2 Lens aperture

Since the wavefront system leaving an object is the only information on which the formation of its image is based it is to be expected, and is indeed found, that the more of the wavefront system that is allowed to enter the

Fig. 1.07 — Arrangements for forming Fraunhofer diffraction patterns.

imaging lens the better is the quality of the image. Rephrased, this is the well-known axiom that the larger the aperture of the lens the better is the definition of the image (assuming aberrations are not a limiting factor).

An alternative model of imaging, that is equivalent to the above, considers the imaging of each point in the object individually. The reader will recognize that this uses the historical work on the resolving power of telescopes, where it will be recalled that the image of a star (a close approximation to a point source) is blurred by diffraction at the lens aperture into a disc surrounded by rings — the '**Airy pattern**' named after Sir George Airy the Astronomer Royal who in 1835 worked out the details of the pattern (§2.3). The dimensions of the Airy pattern are inversely proportional to the diameter of the diffracting aperture, so that only if the aperture were infinitely large would each object point be imaged as a point.

Fig. 1.08 — Two models of image formation.
　　　　　　(i) Wavefronts diffracted from across an object mask form the Fraun-
　　　　　　hofer diffraction pattern of the object in the back focal plane of lens
　　　　　　2, the plane where the image of the light source as object is formed.
　　　　　　The diffracted wavefronts proceed beyond the diffraction plane and
　　　　　　recombine in the image plane with other diffracted wavefronts (not
　　　　　　shown here but see Fig. 5.03) to form an image of the object.
　　　　　　(ii) Light from each object point (P) acts as a source of (diverging)
　　　　　　spherical wavefronts that give the Fraunhofer diffraction pattern of
　　　　　　the lens aperture (the Airy pattern) as the image (P') of that point
　　　　　　source as object.

　　　On this alternative model, each point in an object being imaged is
regarded as a point source of light (Fig. 1.08). The wavefronts from each
object point are diffracted at the lens aperture to give a Fraunhofer
diffraction pattern of that aperture in the plane conjugate to the plane of the
'light source' (cf. final paragraph of the previous section). Each point of the
object is therefore imaged not as a point but as the Airy pattern of the
aperture of the imaging lens (or more complicated pattern if lens aberrations
are present). This is true of any object point no matter where it is.
　　　Both approaches to imaging have their uses, and they are dealt with in
more detail in Chapter 5.

1.4   INTERFERENCE BY DIVISION OF AMPLITUDE

Newton's rings and the fringes seen in thin films such as soap bubbles, oil on
water, etc. are the result of interference occurring when light is partially
reflected at two (or more) successive boundaries between media of different

refractive indices. If a wavetrain of incident light is partially reflected at the first interface (air/oil in the case of oil on water) then a reduced amplitude of the same wavetrain is transmitted and then partially reflected at the next interface (oil/water). Interference occurs if the two reflections are brought together, as when viewed by the eye for example, with a result that depends on the path difference that has been developed between them owing to the separation of the two reflecting surfaces. (In white light, colour effects are observed when the path difference — a function of film thickness and angle of viewing — is such that there is constructive interference for some wavelengths and destructive interference for others.)

In contrast to interference involved in diffraction, where wavefronts are 'divided' by apertures, the above effects are classified as interference by division of amplitude and devices based on this type of interference are 'amplitude-splitting interferometers'. The example depicted in Fig. 1.09(a)

Fig. 1.09 — Interference by division of amplitude.

is of interference between partial reflections at the two surfaces of a thin parallel plate. Each incident wavetrain is partially reflected at the two interfaces, air/glass at O and glass/air at B. If μ is the refractive index of the glass then the optical path difference, l, between the two reflections $(1, 2)$ at O and C is given by

$$l = \mu(OB + BC) - OD$$

Since $\mu OA = OD$ this reduces to

$$l = \mu(AB + BC)$$
$$= \mu BC(\cos 2\theta + 1)$$
$$= \frac{\mu h}{\cos \theta} 2 \cos^2 \theta$$
$$l = 2\mu h \cos \theta \ .$$ (1.12)

Constructive interference between reflections 1 and 2 occurs if

$$2\mu h \cos \theta = (m + \tfrac{1}{2})\lambda$$ (1.13)

where m is an integer or zero. The $\frac{1}{2}\lambda$ on the right side of this equation is to allow for the phase change of π involved in the reflection at O (cf. Stokes's treatment of reflection and refraction (Appendix B)).

If the incident light is not restricted to the plane of the figure concentric circular interference fringes can be observed, as shown schematically in (b) for the light transmitted through the plate. This is because the condition for an intensity maximum at a point such as F is equally fulfilled by all points on a circle through F with its centre on the axis X. (For the transmitted light system the condition for an intensity maximum is $2\mu h \cos \theta = m\lambda$.)

In view of the comments in §1.2 about coherence it is evident that since h can be appreciable, the effect of temporal (rather than spatial) coherence (and thereby the spectral composition of the illumination) on fringe visibility with this arrangement can be considerable. To study this in practice a parallel air-gap of adjustable width is used. In Chapter 7 we meet such an arrangement in the guise of Michelson's spectral interferometer and we see how the variation of fringe visibility with plate separation is related to the spectral composition of the light by a Fourier transform. As mentioned earlier, it forms the basis of some modern methods of spectroscopy.

PROBLEMS

1.1 In Young's experiment, what pinhole separation (D) would give fringes with a spacing of 1 mm on a screen at 1 m from the pinholes, if the wavelength of the light is 500 nm?

1.2 Estimate the useful transverse coherence width on the surface of the

Earth, given by illumination (λ = 550 nm) from the Sun (angular width 0.5°). State the assumptions you make.

1.3 A light source (λ = 550 nm) is to be used to give illumination with a transverse coherence width of 1 mm on a screen. What must be the maximum angular width of the source? State the assumptions you make.

1.4 Show that for the interference fringe system that can be produced by multiple reflections transmitted through a parallel glass plate (Fig. 1.09(b)) the optical path difference between successive transmitted rays (such as those shown in the figure) is $2\mu h \cos\theta$, and that the condition for their reinforcement is given by $2\mu h \cos\theta = m\lambda$.

2

Fraunhofer diffraction

2.1 INTRODUCTION

In this chapter details of Fraunhofer diffraction patterns given by some simple apertures and gratings are described in preparation for subsequent chapters. Much of this should already be familiar to the reader, but certain features of particular relevance later need to be identified and highlighted.

The chapter concludes with a look at the close analogy between optical diffraction and the diffraction of X-rays by crystals. The indirect way in which the 'imaging' of atoms in crystals can be achieved by using X-ray diffraction data, indirect because lenses for X-rays are impracticable, will then be briefly described in Chapter 5 as an application of the theory of image formation.

2.2 SINGLE-SLIT PATTERN

Figure 2.01(a) depicts a section of a slit of width a, and of length l perpendicular to the plane of the figure. The slit is uniformly illuminated with monochromatic light, wavelength λ, fulfilling the requirements of §1.2 for coherence, and the arrangement is such that plane wavefronts arrive at the slit at normal incidence. The Fraunhofer diffraction pattern given by the slit is then formed in the back focal plane of the lens. We assume that $l \gg a$ so that events in planes parallel to that of the figure can be assumed to be the same: such a pattern is a '**1-dimensional**' **diffraction pattern**. Its details can be derived using the Huygens' wavelets model.

Imagine the slit to be divided, parallel to its length, into a large number of narrow strips, all of the same width. The whole slit is uniformly illuminated, the strips are of equal area, so the amplitude of the contribution from each strip in a given direction θ is the same. Calculation of the combination of all the contributions from across the entire slit is most easily done with the help of a phasor diagram. The phase of the light from the central strip at C can serve as a convenient reference baseline in the phasor diagram. In Fig. 2.02(a) the phasor for the light from this strip is labelled C′. The phase of the light from the adjacent strip below C in Fig. 2.01(a) lags behind that from C, and that from the adjacent strip above C is ahead of that from C. In the phasor diagram these are added to C′ accordingly, as are pairs from strips further and further away from C, until the contributions from all the strips have been included. R is the resultant given by this construction, but to obtain an accurate value we need a phasor diagram which represents the limit when the strips become infinitesimally narrow. The transition is easily made. First we note that the present phasor diagram resembles part of a regular polygon because (i) the phasors are of equal length, and (ii) they are rotated through equal angles (owing to the linear variation of phase between the contributions from successive strips). The light from A in Fig. 2.01(a) has a phase ahead of that from C by an amount

$$\alpha = \frac{2\pi}{\lambda} \cdot \frac{a}{2} \sin \theta \tag{2.01}$$

and that from B lags behind that from C by α. For light leaving a position halfway between C and B the phase lag is $\alpha/2$, and so on. In the limit when the strip widths are reduced to zero the phasors form an arc of a circle of radius r, as shown in Fig. 2.02(b), and the resultant amplitude, R_1, is given by

$$R_1 = 2r \sin \alpha \ . \tag{2.02}$$

The subscript is to remind us later that this resultant refers to a single slit.

The arc A′C′B′, of length $r.2\alpha$, is equal to the total length of all the phasor elements stretched out in a straight line, as in (c). Now this straight line represents the phasor diagram for the $\theta = 0$ direction, since in that special direction (cf. Fig. 2.01(a)) there are no path differences and all the contributions are in phase with that from C. We may therefore write

$$R_1(0) = r.2\alpha$$

where $R_1(0)$ denotes the amplitude in direction $\theta = 0$.

Substitution into Eqn (2.02), and noting that R_1 is here a function of θ, we have

$$R_1(\theta) = R_1(0) \left[\frac{\sin \alpha}{\alpha} \right] \ . \tag{2.03}$$

The features of the term in brackets are shown in Fig. 2.03(a).

The experimentally observed intensity pattern is given by

$$I_1 = R_1^2$$

as shown in Fig. 2.03(b) and Fig. 2.12(a).

Fig. 2.01 — Single-slit Fraunhofer diffraction.

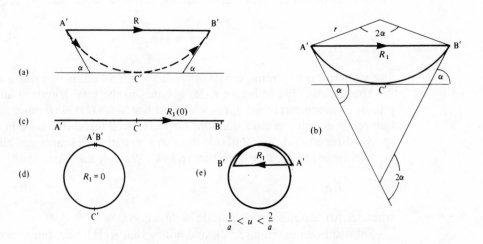

Fig. 2.02 — Single-slit diffraction: phasor diagrams. (The slight displacement in (e) is to show that the overall phase difference is 3π).

The variable α is a function of $(\sin\theta)/\lambda$ and a (cf. Eqn (2.01)). Interpretation of the diffraction pattern is simplified if $(\sin\theta)/\lambda$ is replaced by the single variable

$$u = \frac{\sin\theta}{\lambda}.$$ (2.04)

Fig. 2.03 — Single-slit diffraction pattern. (a) Amplitude (sinc function). (b) Intensity.

Eqn (2.03) can then be written as

$$R_1(u) = R_1(0) \left[\frac{\sin \pi u a}{\pi u a} \right] . \tag{2.05}$$

The bracketed term in this equation is an example of a **sinc function**, of general form $(\sin \pi x)/\pi x = $ 'sinc x', which occurs widely in physics and was given this name by P. M. Woodward in 1953 in his book *Probability and information theory, with applications to radar* (see References).

From Fig. 2.03 there is seen to be a reciprocal relationship between the scale of the pattern in terms of u and the slit width a. Now u has the dimensions of reciprocal length, and we are therefore led to the concept of describing Fraunhofer diffraction patterns as existing in '**reciprocal space**'. This will be developed further when we consider multiple-aperture patterns (§2.4 and §2.5) and it will take on additional meaning when we encounter the Fourier aspects of diffraction (Chapters 3 and 4).

Meanwhile it is necessary to note in Fig. 2.03(a) the variation with u, of the phase of the amplitude pattern. As u starts to depart from zero the resultant amplitude decreases, corresponding to the curving of the phasor diagram (b) in Fig. 2.02. But the resultant remains parallel to the phase baseline, i.e. its phase is that of the illumination in direction $u = 0$. When $u = 1/a$ the phasor diagram has formed a complete circle (d) and the resultant amplitude is zero. [$u = 1/a$ means that $a \sin \theta = \lambda$, i.e. one whole λ path difference between the contributions from A and B (Fig. 2.01(a)). This seems to conflict with a zero nett amplitude from the whole slit, until one notes that it means that there is therefore a $\frac{1}{2}\lambda$ path difference causing cancellation between the contribution from A and C, and similarly between those from just below A and just below C, and so on right across the slit.] After that, as u increases to $u = 2/a$ the phasor digram (Fig. 2.02(e)) exceeds one full circle, and there is again a resultant parallel to the baseline. However, the phase is now π, as manifest in the negative value of the amplitude in Fig. 2.03(a) in this region of u. Further zeros and phase reversals follow at $u = 2/a$, $3/a$, $4/a$, ..., the phasor diagram forming ever-decreasing circles of more and more turns. The condition for the zeros is given by

$$a \sin \theta = m\lambda \ ,$$

where m is an integer.

It is also necessary to note that if the slit in Fig. 2.01 were shifted in its own plane then although the diffraction pattern on the screen would not be displaced (because all parallel beams in a given direction, such as θ in the figure, are brought to the same focus (P)), there would be a corresponding overall change in phase of the pattern, superimposed on the relative phase variation within the pattern as discussed above. The relationship between displacement of the aperture and the change in phase of the diffraction pattern is a linear one and is expressed in the **shift theorem** (§4.3): it applies to any aperture system. However, as the eye is insensitive to phase, the diffraction pattern would *appear* totally unchanged by the displacement of the aperture. Indeed, the effect would only be detected if the diffraction pattern were recorded by a holographic or other technique in which wavefronts from other apertures that have not been moved in the same way are also involved, causing interference effects that would reveal the displacement.

Finally, we can introduce the way in which we shall specify the amplitude distribution of incident illumination across any aperture system — the **amplitude transmittance** of the system. For this, we shall use the conventional '**aperture function**' description (the word 'function' being used in the mathematical sense). In Fig. 2.01 we have said that the slit is uniformly illuminated, and the aperture function then takes the form, shown in (b), of a **rectangle** (or '**top hat**') **function**, $f(x)$, which has a constant value, h, over the width of the slit, and is zero elsewhere. In this chapter we shall be concerned with some other uniformly illuminated aperture systems: the more general case will be encountered in Chapter 5. One can say that, in general, $f(x)\mathrm{d}x$ is the amplitude of wavemotion originating from element $\mathrm{d}x$ at x. The concept is extendable to 2- and 3-dimensions in a completely straightforward way. And in the context of this book the aperture function of any system of apertures (not necessarily just one aperture as here) can usefully be thought of as representing what one can call the (**optical**) '**structure**' of that system.

2.3   CIRCULAR APERTURE

The Fraunhofer diffraction pattern formed by a circular aperture is of considerable importance in connection with the performance of most optical instruments. Unfortunately, the details of the pattern are difficult to derive by the phasor method used in the previous section for a slit aperture. The reason is that the strips into which the aperture is imagined to be divided are not now all the same length (cf. aperture functions in Fig. 2.05(a)). They progressively increase and then decrease in size across the aperture, and the phasor diagram would not form part of a regular polygon. This example is more easily solved analytically and details are to be found in Appendix C. The pattern (Fig. 2.04(a)) consists of a central disc surrounded by concentric

(a)   (b)   (c)

Fig. 2.04 — Circular-aperture diffraction patterns.
(a) Single aperture (the Airy pattern).
(b) Two apertures illuminated separately, incoherently.
(c) Two apertures, illuminated by a common, coherent source.
((a) and (c) reproduced by permission, from Harburn *et al.* (1975), *Atlas of optical transforms*. Bell, London.)

circular fringes and is known as the **Airy pattern**, after Sir George Airy, the Astronomer Royal, who derived the details in 1835.

The amplitude variation along a diameter of the Airy pattern is shown by the full line in Fig. 2.05(b). Though closely resembling the pattern for a slit aperture (shown by the broken line) it is described by a Bessel function instead of the sinc function. For us, an important difference between the two is that the first intensity zero from the central peak of the Airy pattern does not occur until

$$a \sin \theta \approx 1.22\lambda \tag{2.06}$$

where a is the diameter of the aperture. Subsequent zeros are also displaced.

An immediately obvious example of the formation of this pattern occurs when an astronomical telescope is used to image a star, and it was in this context that Airy derived the details of the pattern. A star is effectively at an infinite distance from an observer on Earth and its image is therefore formed in the back focal plane of the telescope objective, where it is examined with the aid of the eyepiece. Since its **angular diameter** (the angle subtended at Earth by its diameter) is extremely small the image should be almost a point. However, the wavefronts from the star are interrupted by the limited aperture of the objective, diffraction occurs and it is the Fraunhofer diffraction pattern of this aperture that is seen as the image of the star. As an example of Fraunhofer diffraction this corresponds to the arrangement

Fig. 2.05 — (a) Aperture functions, $f(x)$: solid line, uniformly illuminated circular aperture of diameter a; dashed line, uniformly illuminated slit of width a.

         (b) Amplitude diffraction patterns: solid line, circular aperture of diameter a; dashed line, slit of width a.

         (c) Rayleigh criterion (circular aperture patterns).

shown in Fig. 1.06, with the aperture of the objective itself acting as the diffracting mask.

No matter how free from imperfections an astronomical telescope objective (lens or mirror) is, what is observed at best is not a point image of the star, but the Airy intensity pattern produced by the aperture of the telescope objective. In the wider context of Chapter 5 this pattern — the response of the system to a point ('impulse') input (the **impulse response**) — is the **point-spread function** (**PSF**) of the system.

The optical case we have just considered (where the image that an aberration-free astronomical telescope produces of a star is not a point but is the Airy pattern of the aperture of the telescope objective) is an example of what is described as a '**diffraction limited**' system. It means that the performance of the system is only limited by diffraction, due to the finite sizes of its optical components.

It is worth noting here the wider implications of this. For example, to have a truly '**parallel beam**' of light the transverse dimensions of the beam would need to be infinite. Of course, unless the restricting apertures are very small the illumination may well be effectively parallel for all practical purposes. We can appreciate this by noting from Fig. 2.05 that the angular width of a beam from a circular aperture — taking the **angular width** as the full width of the **Airy disc** at half maximum intensity — is given by $\Delta\theta \approx 1.22\lambda/a$. Calculation of $\Delta\theta$ for various values of a will show the truth of the above remark. In the same way, a beam of coherent light from a **laser** spreads due to its finite width as it leaves the laser.

From an imaging point of view, it is therefore the diameter of the objective of a telescope, and the wavelength of the light, that determines the **limiting resolution** of the telescope when aberrations have been eliminated or minimized.

Consider the image of two stars of small separation θ (Fig. 2.06). As they

Fig. 2.06 — The astronomical telescope. The Rayleigh criterion applied to the imaging of two stars of small angular-separation.

are mutually incoherent sources the image consists of two independent Airy intensity patterns. Resolving the presence of two stars depends, therefore, on the sizes of the Airy discs and the extent to which they overlap. The conventional specification for the limiting condition, the **Rayleigh criterion**, is the separation shown in the above figure and also in Figs. 2.04(b) and 2.05(c). According to this the two patterns are resolved if the centre of the Airy disc of one pattern is superimposed on the first dark ring from the centre of the other. This gives a dip in the total overlap intensity of about 20% between the two peaks (assumed equal in magnitude for the present purpose). A dip of this magnitude, though rather arbitrary, is a convenient criterion of resolution for many pourposes.

On the basis of the Rayleigh criterion, Eqn (2.06) then gives the angular **resolution limit**, θ, for a **telescope** of aperture a as

$$\theta \approx 1.22\lambda/a \tag{2.07}$$

or

$$\theta \approx \frac{0.15}{a_{\text{metres}}} \text{ secs of arc}$$

as a rough guide for yellow light.

(To try to improve the ability of a telescope to resolve very close images of stars, the circular fringes surrounding an Airy disc can be suppressed. This is done by '**apodization**' — modifying the aperture function of the objective by using, for example, a glass plate whose transmission properties vary from point to point in a prescribed way. The price paid for restricting the transmission of light through the objective is, however, liable to be the broadening of the actual disc itself.)

The same factors determine the resolution limit of a terrestrial telescope or camera. Under normal conditions of illumination every point of a terrestrial object scatters light independently of its neighbours and is therefore independently imaged. This is arguably the same as if a cluster of stars were being imaged. For this reason the term '**self-luminous object**' tends to be used with some licence in both contexts as a short way of referring to objects imaged under incoherent conditions. With a terrestrial telescope or camera, the image corresponding to every object point as source again consists not of a point but of the diffraction pattern of the aperture of the objective (cf. §1.3.1). (We shall not consider the role of the **eyepiece** in image formation with a telescope — or compound microscope (below) — since it is a relatively low-powered second-stage component and not a prime source of imperfections.)

A similar situation applies when an optical **microscope** (simple or compound) is used to examine a 'self-luminous' object under conditions like those considered above. Using the Rayleigh criterion as for the telescope, the minimum separation s in Fig. 2.07(a), of two points in an object that can

Fig. 2.07 — The optical microscope (self-luminous object).
     (a) The Airy-pattern images of two object points O and O′, distance s apart, are resolved according to the Rayleigh criterion, by the objective lens, if $O'B - O'A = 1.22\lambda$, since $OA = OB$. From (b): $O'B = OB + s\sin i$, $O'A = OA - s\sin i$, whence $s = 0.6\lambda/\sin i$.

be resolved by a microscope, is given by

$$s \approx 0.6\lambda/\sin i \qquad\qquad (2.08)$$

where i is the semi-angle subtended by the objective at the object. If the object space is filled with a medium of refractive index μ — using an 'oil-immersion objective' — then the wavelength of light there is reduced to λ/μ and the above expression becomes

$$s \approx 0.6\lambda/\mu\sin i \qquad\qquad (2.09)$$

where $\mu \sin i$ is the **numerical aperture** of the microscope objective. The maximum possible value of i is 90°, giving the '**microscopic limit**' as approximately $\frac{1}{2}\lambda/\mu$.

With optical imaging systems in general there is usually a limiting aperture that controls the light-gathering power of the system. The '**aperture stop**', often located between the various lens components of the systems, inevitably gives rise to diffraction as described above. As seen from the object (i.e. source) this aperture is called the **entrance pupil**, and as seen from the image it is called the **exit pupil**. In the language of instrumental optics the pupils are thus the images of the aperture stop as formed in the object and image spaces. And the aperture function, already defined in §2.2, when expressed in terms of the image-space coordinate system, is called the **(exit) pupil function**.

2.4 DOUBLE APERTURE

2.4.1 Two slits

The derivation of the details of the Fraunhofer diffraction pattern obtained with a double-slit arrangement provides a useful introduction to the multiple-slit pattern (§2.5), where exactly the same principles are involved.

The arrangement shown in Fig. 2.08(a) is similar to that for the single slit,

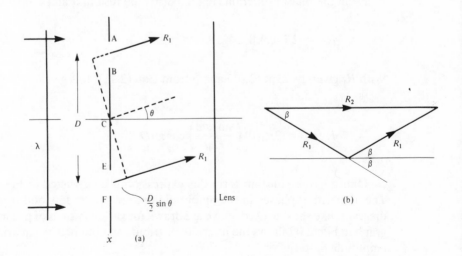

Fig. 2.08 — Double-slit diffraction.

but with two slits parallel to each other and with their centre-lines distance D apart. The spatial pattern of complex amplitude leaving each slit is exactly the same as before but the nett result in any direction θ depends on the path difference between the two contributions in that direction. For example, only when θ is such that the diffracted light from the two slits is in step is there complete reinforcement. From the figure it follows that this condition is fulfilled when

$$D \sin \theta = n\lambda$$

i.e. $\quad \dfrac{\sin \theta}{\lambda} = u = \dfrac{n}{D} \quad \text{(using 2.04)}$ $\qquad$ (2.10)

where n, the **diffraction order**, is zero or an integer. In all other directions there is destructive interference to different extents.

A phasor diagram leads to an expression for the whole pattern. Relative to a phase reference origin chosen for convenience at C in the figure, slit AB diffracts in direction θ with a phase $+\beta$ and slit EF diffracts with a phase $-\beta$, where

$$\beta = \frac{2\pi}{\lambda} \frac{D}{2} \sin \theta$$

$\qquad$ (2.11)

i.e. $\quad \beta = \pi u D$.

From the phasor diagram (Fig. 2.08(b)) the resultant amplitude is

$$R_2 = 2R_1 \cos \beta .$$

With R_1 given by Eqn (2.05) and β from Eqn (2.11), this gives

$$R_2(u) = 2R_1(0) \left(\frac{\sin \pi u a}{\pi u a} \right) \cos \pi u D .$$

$\qquad$ (2.12)

Omitting the constant term this expression is constructed in Fig. 2.09. The alternating phases in the amplitude pattern are understood if phasor diagrams like the one used above are drawn for various values of β. The final graph in Fig. 2.09 shows the intensity distribution, obtained by squaring the amplitude pattern, viz.

$$I_2(u) = 4I_1(0) \left(\frac{\sin \pi u a}{\pi u a} \right)^2 \cos^2 \pi u D .$$

$\qquad$ (2.13)

An example of the observed pattern in shown in Fig. 2.12(b).

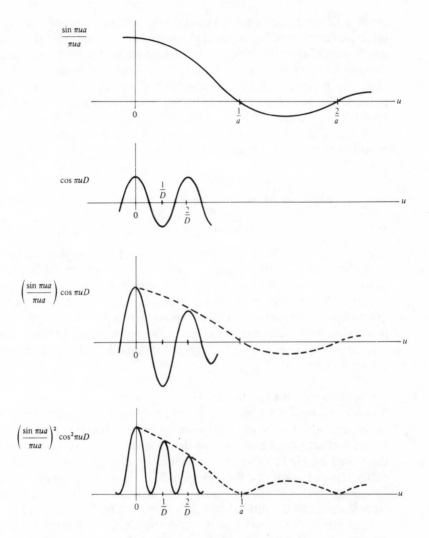

Fig. 2.09 — Double-slit diffraction pattern.

There are several important points to note in connection with the double-slit pattern. As Eqn (2.13) and Fig. 2.09 show, the pattern is the product of two diffraction 'terms': the single-slit pattern is multiplied by a $\cos^2$ fringes pattern associated with interference between the diffracted light from the two slits.

The maxima of the $\cos^2$ fringes correspond to directions in which there is constructive interference between light from the two apertures. We then have $D \sin \theta = n\lambda$, i.e. $u = n/D$ (Eqn (2.10)), the $\cos^2$ term is unity, and the observed intensity is — as one expects — the square of the sum of the separate amplitudes from the two apertures. In a similar way, between these directions there is destructive interference, the $\cos^2$ term is zero and the observed illumination is zero. Thus the observed maxima and minima are in directions determined by the aperture separation D. Note that their intensities, however, are determined by the amplitudes of the light diffracted from the apertures in those directions. In this sense the observed diffraction pattern can be regarded as an 'amplified sampling' of the single-aperture pattern, the sampling being restricted to directions determined by the aperture separation. This is evident in Figs. 2.09 and 2.12(b) where the single-slit term is seen to be the envelope of the $\cos^2$ fringes. Inserting the condition $u = n/D$ for reinforcement into Eqn (2.13) gives

$$(I_2)_{max} = 4I_1(0) \left[\frac{\sin \frac{\pi na}{D}}{\frac{\pi na}{D}} \right]^2 \tag{2.14}$$

and the term in brackets is what can be described as a **sampled sinc function**.

In accordance with the interpretation of u as a parameter in reciprocal space (§2.12), this sampling of the single-aperture pattern can be regarded as taking place in that space. We can say that the double-aperture pattern is a sampling in reciprocal space of the single-aperture pattern; and the sampling is at values of u determined by, and reciprocally related to, the aperture separation D, i.e. at $u = n/D$. Thus the smaller the value of D the wider is the separation of the fringes.

2.4.2 Two circular apertures

The derivation of the diffraction pattern produced by two circular apertures distance D apart is closely analogous to that from two slits: the sampling aspect is exactly the same. The full details need not be worked out here. As illustrated in Fig. 2.04(c) we again have the single-aperture diffraction pattern (here the Airy pattern) multiplied by the same $\cos^2$ term as above — i.e. the term that is associated with the separation (D) of two apertures. Note in Fig. 2.04 the difference between (b) and (c). It will be recalled that the pattern in (b) consists of two quite separate Airy **intensity** patterns, produced incoherently by different sources. In (c), where there is coherence, it is the summation of the two Airy **amplitude** patterns that is involved and the $\cos^2$ sampling fringes are in evidence.

2.5 N-SLIT GRATING

As with the double slit, complete reinforcement between the light leaving successive slits in an N-slit grating only occurs when $D \sin \theta = n\lambda$, where D is here the **grating repeat** and n the grating **diffraction order**. (Other, subsidiary maxima do occur with a multiple aperture grating (see, for example, Fig. 2.12) but when N is large these are negligible for most purposes.) The (**principal**) **diffraction maxima** are, as for the double slit; an amplified ($N \times$) sampling of the single-slit pattern: the sampling is in the directions given by $D \sin \theta = n\lambda$, and the amplitudes are $N \times$ the amplitude from each individual slit in those directions.

For our needs here, the reader can therefore proceed directly to Eqn (2.17), but for completeness the fuller details of the N-slit pattern are given below.

The same phasor method as for the double slit can be used. For N slits there are N phasors (Fig. 2.10) each of length R_1, this being the amplitude from

Fig. 2.10 — N-slit diffraction.

each individual slit as before (see Eqn (2.05)). Again, there is an angle of 2β between successive phasors, where $\beta = \pi u D$ (Eqn (2.11)). From Fig. 2.10 we see that the ends of the phasors lie on a circle of radius $r' = R_1/2\sin\beta$. The resultant, R_N, for N slits is therefore

$$R_N = 2r'\sin N\beta = R_1 \frac{\sin N\beta}{\sin\beta} \ .$$

Substituting for β gives

$$R_N = R_1 \frac{\sin N\pi u D}{\sin \pi u D}$$

where D, the distance between the centres of adjacent slits is the **'grating repeat'**.

Using Eqn (2.05) for R_1 as before, and reminding ourselves that R_N is a function of u, we have

$$R_N(u) = R_1(0)\left(\frac{\sin \pi u a}{\pi u a}\right)\left(\frac{\sin N\pi u D}{\sin \pi u D}\right) \ . \tag{2.15}$$

The intensity is given by the square of this expression, viz.

$$I_N(u) = I_1(0)\left(\frac{\sin \pi u a}{\pi u a}\right)^2\left(\frac{\sin N\pi u D}{\sin \pi u D}\right)^2 \tag{2.16}$$

Again, we have the (squared) product of the diffraction 'terms', the **single-slit term** and what is here the **'grating term'** — becoming the $\cos^2 \pi u D$ term in Eqn (2.13) when $N = 2$.

Figure 2.11 represents a graphical construction of the grating term. The

Fig. 2.11 — N-slit diffraction pattern.

main peaks are the grating **principal maxima** referred to at the beginning of this section. They occur when there is full reinforcement between the diffracted light from successive slits and, as for the double slit, this happens when

$$D \sin \theta = n\lambda \ ,$$

i.e.
$$u = n/D \ ,$$

where n is the **grating diffraction order** (cf. Eqn (2.10)).

In the directions of the grating principal maxima the value of the grating term is N^2, because as $u \to n/D$ we have

$$\left(\frac{\sin N\pi u D}{\sin \pi u D}\right)^2 \to \left(\frac{N\pi u D}{\pi u D}\right)^2 = N^2 \ .$$

Since N is large for the type of grating normally used — usually tens of thousands — the principal maxima and the weaker, **subsidiary** (secondary) **maxima** become very sharp, as can be appreciated in Fig. 2.11. The figure also shows that when N is large the subsidiary maxima are close together. The overall effect on the subsidiary maxima is that they become insignificant in the pattern; we shall ignore them hereafter.

The intensity of illumination in the directions of the principal maxima is given by the square of the product of N and the amplitude of diffraction from each slit in those directions. For those directions Eqn (2.16) becomes the **amplified sampled sinc² function**, as for the double slit ($N = 2$), namely

$$(I_N)_{max} = I_1(0) \left[\frac{\sin \dfrac{\pi na}{D}}{\dfrac{\pi na}{D}} \right]^2 N^2 . \tag{2.17}$$

Figure 2.12 shows the patterns given by gratings with N up to 6. The way

Fig. 2.12 — Diffraction patterns given by various numbers of slits. (Reproduced from *Studies in Optics* by A. A. Michelson by permission of the University of Chicago Press. © 1927 by the University of Chicago.)

in which principal maxima are a sampling of the single-slit pattern is demonstrated. Note too, how there are '**missing orders**' if the amplitude from the individual slits happens to be zero in those directions. We have seen that (1) a grating diffraction maxima occurs in directions given by

$$D \sin \theta = n\lambda$$

where n is the 'order', and that (2) the amplitude from an individual slit is zero in directions given by

$$a \sin \theta = m\lambda$$

where m is an integer (§2.2). The directions for these two conditions therefore coincide if

$$n = mD/a .$$

Thus, for example, if $D/a = 4$ (as in Fig. 2.12) then the orders $n = 4, 8, 12$, ..., are missing.

The single-slit term and the grating term may both be described as existing in reciprocal space (§2.2). Indeed the spatial distribution of the principal maxima in the Fraunhofer diffraction plane can be described as the reciprocal lattice of the grating — albeit a 1-dimensional lattice in the present instance. Thus we can say that for large N, when the principal maxima are sharp, the pattern can be regarded as an amplified sampling of the single-slit pattern at the reciprocal-lattice points of the grating. This is an important concept, applicable to any type of grating.

Finally, it should be noted that as the pattern is symmetrical about the central (zero order) maximum, all the other maxima (meaning now the principal maxima) are symmetrically spaced in pairs, of diffraction order $\pm n$, about the central maximum.

2.6 2-DIMENSIONAL GRATINGS

No new principles are involved in extending the findings of the previous sections to two dimensions. The only result that we need to note here is summarized diagrammatically in Fig. 2.13. An aperture mask containing a

Fig. 2.13 — 2-dimensional diffraction.

large number of pinholes arranged to form a 2-dimensional lattice, of which just a small part is represented in (a), gives a diffraction pattern consisting of a lattice of spots arranged as indicated in (b). Consistent with previous results, the scale of the diffraction pattern is reciprocally related to that of the pinhole array and rows of spots in the former are perpendicular to rows of pinholes in the latter. If $\mathbf{a}, \mathbf{b}$ are the translation vectors of the grating lattice in (a), and $\mathbf{a}^*, \mathbf{b}^*$ those of the diffraction pattern in (b) we have

$$|\mathbf{a}^*| = K/|\mathbf{a}| \ , \qquad |\mathbf{b}^*| = K/|\mathbf{b}|$$

where K is a constant. Also we have

$$\mathbf{a}^* \cdot \mathbf{b} = 0 \ , \qquad \mathbf{b}^* \cdot \mathbf{a} = 0 \ .$$

In keeping with the reciprocity, a mask with pinholes arranged like the spots in (b) would give a diffraction pattern like the arrangement in (a). (It is a convention to use * to denote reciprocal lattice dimensions and it does not refer to the complex conjugate here.)

2.7 CRYSTALS AS 3-DIMENSIONAL GRATINGS

In crystals a group of atoms (sometimes — as in some metals — just one atom) is associated identically with each point of a regular space-lattice. The lattice repeats are generally between about 1 nm and 10 nm, though in crystals of biological molecules, for example, they are larger than this because so many atoms are associated with each lattice point. Since X-rays have wavelengths in the range from about 10^{-2} nm to about 10^2 nm (Appendix K) and are scattered (or diffracted — the terms are used interchangeably as noted in §1.3) by the electron clouds in atoms, crystals act like 3-dimensional diffraction gratings to X-rays. The **'Laue experiment'** in 1912 demonstrated the reality of this, suggesting that contrary to some views at that time, X-rays are waves, or at least have wavelike properties. In 1914 Max von Laue was awarded the Nobel Prize in physics for this. In the following year the prize was awarded to W. L. Bragg and his father W. H. Bragg for their immediate application of the diffraction effects to the analysis of crystal structure by means of X-rays — now the subject of **X-ray crystallography**. As we have already noted, atoms cannot be imaged directly by focusing X-rays and the process has to be completed indirectly from the information provided by the diffraction effects.

Because a crystal is like a 3-dimensional grating, not just 1- or 2-dimensional, the conditions necessary for the equivalent of optical principal diffraction maxima to occur are not easily fulfilled. Consider the unit cell of a crystal lattice as depicted in Fig. 2.14(a). Imagine a train of quasimonochromatic waves of wavelength λ to sweep through the crystal. What is the basic requirement for obtaining a diffraction maximum in some direction? It is that the X-ray scattering in that direction by the (identical) atomic assemblages centred at lattice points A, B, and C must be in phase with that

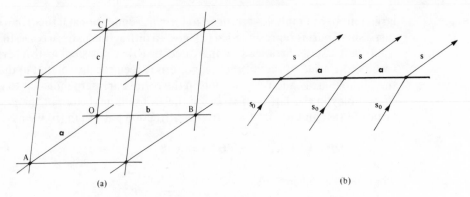

Fig. 2.14 — A crystal lattice as a 3-dimensional diffraction grating.

scattered from the assemblage at O. The waves scattered from these centres will then be in phase with the scattering from their neighbours, and so on throughout the crystal. The fact that not even one atom need be located actually at a lattice point does not affect this requirement for a diffraction maximum because the lattice repeat — the distance between corresponding atoms located identically in relation to successive lattice points — is all that matters. The lattice point is hypothetical, and its location with respect to the atomic assemblage that it represents is arbitrary, though other considerations, such as symmetry, influence its choice. On the other hand, the *intensity* of a diffraction maximum is affected by the mutual positions of the atoms within the assemblage, and we return to this later in this section.

Consider the repetition of lattice points along the **a**-direction in the crystal (Fig. 2.14(b)). For X-rays incident in a direction specified by unit vector $\mathbf{s}_0$ the nett scattering from the assemblage at one lattice point, in a direction specified by unit vector **s**, will be in phase with that from the next if

$$\mathbf{a}\cdot\mathbf{s} - \mathbf{a}\cdot\mathbf{s}_0 = (\mathbf{s}-\mathbf{s}_0)\cdot\mathbf{a} = h\lambda$$

where h is an integer or zero and **a** is the lattice translation vector. Neither **s** nor $\mathbf{s}_0$ need be in the plane of the figure.

For the complete 3-dimensional array three conditions must simultaneously be fulfilled, viz.:

$$\left. \begin{array}{l} (\mathbf{s}-\mathbf{s}_0)\cdot\mathbf{a} = h\lambda \\ (\mathbf{s}-\mathbf{s}_0)\cdot\mathbf{b} = k\lambda \\ (\mathbf{s}-\mathbf{s}_0)\cdot\mathbf{c} = l\lambda \end{array} \right\} \quad \textbf{Laue equations (or conditions)} \qquad (2.18)$$

where h, k, l are integers or zero, and **a**, **b**, **c** are the translation vectors of the lattice. Note that the lattice need not be orthogonal.

With a crystal at a fixed orientation to the direction of an incident X-ray beam it is unlikely that the above conditions can be fulfilled for more than a few diffraction maxima unless — as in the original experiment — a continuous spectrum of X-ray wavelengths is used. Not knowing what wavelength is responsible for any particular diffraction maximum has obvious disadvantages however. Except for certain types of investigation the common practice now is to use quasimonochromatic radiation and to change the inclination of the crystal systematically during exposure to the X-ray beam to enable the Laue conditions to be met. Details of how this is done need not concern us here. Suffice it to say that it is possible to obtain the 3-dimensional diffraction pattern given by a crystal. It forms a 3-dimensional lattice that is reciprocal to that of the crystal, as with the 2-dimensional gratings considered in the previous section.

The *directions* of the X-ray diffraction maxima from a crystal are determined, for a given wavelength of X-rays, by the dimensions of the crystal grating. By measuring those directions the lattice geometry can therefore be determined. Similarly, the *intensities* of the diffraction maxima are determined by the intensity of scattering, in the directions of the maxima, from the atomic group associated identically with each lattice point in the crystal. Those intensities therefore contain information about the composition and arrangement of the atoms in the group associated with each lattice point.

All this is directly analogous to the way in which the optical diffraction-grating maxima are an amplified sampling of the diffraction from one grating aperture alone. And the pattern from one aperture is determined by its aperture function — the optical structure of the aperture.

We are indeed fortunate to be able to gather this amplified sampling of the scattering of X-rays from the unit cell of a crystal. Even if one unit cell of a crystal could be isolated and handled, its scattering of X-rays would be far too weak to measure. Instead, we have the enormously amplified signals from a real crystal. The price we pay is that these signals are restricted to certain directions determined by the crystal lattice geometry. However, the number of directions is adequate, and it is possible to construct a very detailed picture of the atomic arrangement and electron density distribution in a crystal structure. (The diffraction of X-rays by non-crystalline materials such as glasses and liquids also reveals information about their structure, but details of this would take us beyond the scope of this book.)

In Chapter 5 we shall see how the use of X-ray diffraction data to determine the arrangement of atoms in crystals is, in essence, an application of the Abbe–Porter theory of optical image formation. The close analogy between the principles involved in the imaging of crystal structure by X-rays and conventional optical imaging was a topic of almost life-long interest to Sir Lawrence Bragg, a topic to which he made important contributions over many years.

In connection with the Fourier aspects that we are particularly concerned with it will be useful to derive an expression for the amplitude of an X-ray diffraction maximum from the unit cell of a crystal. This can be done by using the Bragg reflection concept, with which the reader should be familiar.

(It is shown in Appendix E that a diffraction maximum corresponding to a particular set of values of h, k, l in the Laue equations may be regarded as a reflection of the incident X-ray beam by lattice planes in the crystal, defined by the same h, k, l values.)

Figure 2.15 represents an edge-on view of parts of a set of lattice planes

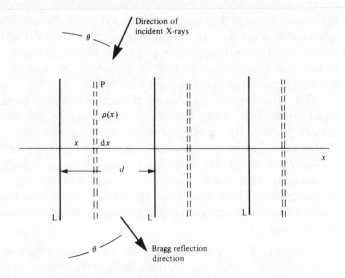

Fig. 2.15 — Bragg reflection. L: edge-on view of lattice plane.

of spacing d. X-rays are incident to the planes at the Bragg angle, θ, for the X-ray wavelength used. Thinking in terms of 'point atoms' for the moment, atoms in any one lattice plane scatter in phase in direction θ: this is the essence of 'reflection'. For reflections from successive lattice planes to be in phase the requirement of the Bragg equation (Appendix E) has to be met, namely

$$2d\sin\theta = n\lambda \tag{2.19}$$

where n is the order of reflection.

As we have already noted, however, atoms are not point scatterers, nor are they confined to lattice points. It is necessary to think of a continuous distribution of scattering matter, that repeats regularly throughout a crystal. (The symmetry aspects of a crystal structure need not distract us here.) Consider, with the aid of Fig. 2.15, the first-order reflection ($n = 1$) associated with the lattice spacing d. The scattering from the part of the electron-density distribution actually in the lattice planes is all in phase in the direction of the reflection, as we have noted above: but we must take into account the scattering from the matter between these planes. For the first-order reflection there is a path difference of one wavelength (a phase difference of 2π) in the scattering from successive lattice-planes. The contribution to the same reflection from the scattering matter in plane P is therefore out of phase by $(x/d)2\pi$. If $\rho(x)$ is the amplitude of scattering from a layer of thickness dx at x, then the total amplitude of this particular reflection from all the scattering matter in one lattice repeat of the structure is given by

$$A_1(\theta) = \int_0^d \rho(x) e^{i(x/d)2\pi} dx \ . \tag{2.20}$$

For the second-order reflection there are two wavelengths path difference in the reflections from successive planes and the phase difference for plane P will then be $(2x/d)2\pi$. Similarly, for the nth order reflection the phase will be $(nx/d)2\pi$ and the corresponding equation becomes

$$A_n(\theta) = \int_0^d \rho(x) e^{i(nx/d)2\pi} dx \ , \tag{2.21}$$

where θ is related to d by the Bragg equation (2.19).

'Reflection' is of course possible at each side of a set of planes and the amplitudes are usually equal (Friedel's law). One can therefore regard this equation as representing the amplitude of pairs of reflections, or equally, the amplitude of pairs of diffraction maxima, of order $\pm n$.

PROBLEMS

2.1   A parallel beam of light of wavelength 546 nm is incident normally at a slit of width 1 mm. A lens is mounted behind the slit and focused on a screen located in its focal plane, 100 cm away. At what distance, x (in cm) from the centre of the diffraction pattern of the slit is the third minimum?

2.2   If the parallel beam of light used in Problem 2.1 is produced by a slit source at the focus of a lens of focal length 20 cm, how wide could this slit source be before the quality of the pattern deteriorates?

2.3   A helium-neon laser emits a diffraction-limited beam ($\lambda = 633$ nm) of diameter 2 mm. What diameter of light patch would the beam produce on the surface of the Moon at a distance of 376.10^3 km from the Earth? Thence explain the use of a large reflector as beam expander, in lunar ranging experiments.

2.4   If a diffraction-limited laser beam ($\lambda = 633$ nm) of diameter 5 mm is directed at the Earth from a space laboratory orbiting at an altitude of 500 km, how large an area (km^2) would the central beam illuminate? State the assumptions you make.

2.5 Monochromatic, coherent illumination (λ) is incident normally at two slits each of width $a = 0.17\,\text{mm}$ and separation $D = 0.85\,\text{mm}$ (distance between *centres*). Which Fraunhofer diffraction orders are 'missing'? What is the intensity of the third order, relative to the central (zero) order?

2.6 Show that if the repeat, D, of a grating has an aperture function $A\sin(2\pi mx/D)$, where A is a constant and m is an integer, then the diffraction pattern given by the grating is restricted to the pair of diffraction maxima of order $n = m$.

2.7 Use a phasor diagram to determine the diffraction angle, θ, at which the central (zero order) intensity maximum from a 6-slit grating, $D = 6\lambda$ and $a = \lambda$, drops to zero.

Calculate the diffraction angles of the first and second order maxima, and their relative intensities.

3

Fourier series and periodic structures

Fourier's theorem is not only one of the most beautiful results of modern analysis,
but it may be said to furnish an indispensable instrument in the treatment of nearly
every recondite question in modern physics.

attrib. Lord Kelvin

3.1 INTRODUCTION

In Chapter 2 we have seen, in a rather restricted way, how diffraction data
given by a periodic object in the form of an optical grating are determined by
the 'structure' of the grating as specified by its aperture function; and that
the same is true of the X-ray diffraction data obtained from the atomic
arrangement comprising the periodic grating-like structure of a crystal. We
have also noted that optical diffraction is an intermediate step when images
are formed with a lens, the lens performing the task of recombining the
diffracted light, in the image plane. With X-rays the action of a lens is not
available, and to form an image of the structural arrangement of atoms in a
crystal with X-rays other means have to be used to complete the imaging
process beyond the diffraction stage.

For a deeper understanding of these matters, and indeed for other
reasons that are described in Chapter 5, we need to know more about the
relationship between a diffraction pattern and the structure of the object
giving it. This is where Fourier methods start to play their vital role.

Continuing, for the time being, to restrict ourselves to periodic objects,
such as multiple-aperture gratings or crystals, the overall aperture function
(or 'structure') of these can be built-up mathematically by the summation of
an infinite series of sinusoidal harmonics — a **Fourier series**, named after
J. B. J. Fourier, the pioneer of the mathematical technique. This statement
that a periodic function can be expressed as the sum of an infinite series of
harmonics is known as **Fourier's theorem.** Most functions obey it, but

details of exceptions — though none are encountered in this book — will be found in mathematics textbooks.

Strictly, the pattern that can be represented by a Fourier series should be repeated an infinite number of times. However, for the applications we shall meet the number of repeats will be sufficient for the use of the series to be justified. In most practical gratings the number of repeats is very large, and in a crystal of 1 mm linear dimension there are more than a million repetitions of the atomic 'pattern unit'.

In this chapter we shall witness the extraordinary fact that the diffraction pattern given by a grating is a statement, almost a physical manifestation, of the Fourier harmonics that comprise the **(overall) aperture function** of the grating (§3.4.1).

Since diffraction is an intermediate step in optical image formation, one consequence is that at the diffraction stage we can control in a calculated way the formation of an image; this is the basis of many aspects of optical processing (Chapter 5). Another major consequence has been the development of methods for the determination of the atomic structure in crystals, no matter how complex, from their X-ray diffraction data.

In Chapter 4 the relationships concerning periodic structures are extended to ones that are non-periodic. But first, some details about Fourier series (§3.2 and §3.3) and their relationship to diffraction by periodic structures (§3.4).

3.2 FOURIER SERIES

The use of Fourier series for representing a periodic pattern such as the optical structure of a grating, can be illustrated by considering the pattern, of which only two units are shown, in Fig. 3.01(a). This particular shape is

Fig. 3.01 — Fourier series representation of a periodic function $f(x)$.

chosen because it shows rather clearly how it can be represented by a Fourier series, and also because of the way it can represent an interesting example later (§3.4.2).

A Fourier series to represent a 1-dimensional periodic function, $f(x)$, can be written

$$F(x) \ = \ \frac{A_0}{2} + A_1 \cos\left(\frac{2\pi x}{D} - \alpha_1\right) + A_2 \cos\left(\frac{2\pi x}{D/2} - \alpha_2\right) + \ldots$$

$$= \ \frac{A_0}{2} + \sum_{n=1}^{\infty} A_n \cos\left(\frac{2\pi n x}{D} - \alpha_n\right) \tag{3.01}$$

i.e. an infinite series of harmonic 'waves' with amplitudes A_n (the Fourier **coefficients**), phases α_n, and progressively decreasing 'wavelengths' D/n — hereafter called 'repeats' to avoid confusion with optical wavelengths. $A_0/2$ is a constant, the factor 1/2 serving only to simplify the writing of some equations later.

An analytical procedure for determining the coefficients for any given $f(x)$ is described in §3.3. Meanwhile, Fig. 3.01(b) illustrates graphically how the first few terms in the **Fourier analysis** of the pattern can at least be recognized qualitatively. In this figure we see that the term for $n = 1$, i.e. $A_1 \cos((2\pi x/D) - \alpha_1)$, has some similarity to $f(x)$ if $\alpha_1 = \pi$. The value of α_1 can only be 0 or π because $f(x)$ in this example is an 'even' function and $\alpha = 0$ would make the resemblance worse. At least the high peaks in the pattern are roughly represented but not the weak ones. The term for $n = 2$, with $\alpha_2 = 0$, helps considerably. Added to the $n = 1$ term with a suitable value of its amplitude A_2, it starts to establish the weak peaks and to enhance the strong ones. Addition of higher harmonics progressively sharpens the match. So far, however, the summation is equally distributed above and below the x-axis. It is the constant term $A_0/2$ in Eqn (3.01) that raises the pattern up from the x-axis to the actual level of the $f(x)$ curve.

The α_n in Eqn (3.01) are the **phases** of the harmonics. For example, phase $\alpha_1 = \pi$ means that the cosine term for $n = 1$ is shifted along the x-axis by $D/2$ compared to a cosine of zero phase.

The summation of the harmonics, the **Fourier synthesis**, is formally to $n = \infty$. In practice, however, fewer than a hundred terms are often found to be quite adequate, though of course in some instances many more are needed.

In a similar way to that for the pattern we have just considered, the structure of an optical diffraction grating can be represented by a Fourier series. For example, as an extension of the notion of an 'aperture function' of a single slit (§2.2) the (**overall**) **aperture function** of a uniformly illuminated **multiple-slit grating** can be represented by the function $f(x)$ in Fig. 3.03(a). The Fourier series to represent this, and the extremely important relationship between the series and the diffraction pattern given by the grating, are dealt with in §3.4. To that end, the method used for the quantitative determination of the terms in a Fourier series is established in the next section.

In connection with our use of Fourier series in diffraction, it is particularly helpful to think of the frequencies of the harmonics as well as their repeats. Just as a temporal periodicity T has a temporal frequency $1/T$, so does a spatial harmonic of repeat D/n have a **spatial frequency** of n/D; it is the number of repeats in one unit of pattern.

3.3 DETERMINING FOURIER COEFFICIENTS: EVEN FUNCTIONS

For the even functions considered in the previous section, Eqn (3.01) can be written as

$$f(x) = \frac{A_0}{2} + \sum_{n=1}^{\infty} A_n \cos \frac{2\pi n x}{D} , \qquad (3.02)$$

if it is understood that A_n is not necessarily positive.

Multiplying both sides by $\cos(2\pi h x/D)$, where $h = 0$ or an integer, and integrating over the repeat D, we have

$$\int_D f(x) \cos \frac{2\pi h x}{D} \, dx = \frac{A_0}{2} \int_D \cos \frac{2\pi h x}{D} \, dx + \int_D \sum_{n=1}^{\infty} A_n \cos \frac{2\pi n x}{D} \cos \frac{2\pi h x}{D} \, dx$$

$$= \frac{A_0}{2} \int_D \cos \frac{2\pi h x}{D} \, dx + \frac{1}{2} \left[\int_D \sum_{n=1}^{\infty} A_n \cos \frac{2\pi (n+h) x}{D} \, dx + \int_D \sum_{n=1}^{\infty} A_n \cos \frac{2\pi (n-h) x}{D} \, dx \right] .$$

The integrand of every term in the first summation on the right side of this equation is zero, since n is an integer and $h = 0$ or an integer. The same applies to the second summation except for the one term where $h = n$,

which is $\int_D A_n \cos 0 \, dx = A_n D$. The first term on the right side is also zero for $h = n \neq 0$, so the whole equation reduces to

$$\int_D f(x) \cos \frac{2\pi nx}{D} dx = \frac{1}{2} \Big[0 + A_n D \Big]$$

whence
$$A_{n \neq 0} = \frac{2}{D} \int_D f(x) \cos \frac{2\pi nx}{D} dx \ .$$

When $h = 0$ only the first term is non-zero, and this gives

$$\int_D f(x) \, dx = \frac{1}{2} A_0 D$$

whence
$$A_0 = \frac{2}{D} \int_D f(x) \, dx \ .$$

The reason for the arbitrary factor of $\frac{1}{2}$ in the constant term in Eqn (3.01) is now evident: it enables the above two expressions for the coefficients to be combined as

$$A_n = \frac{2}{D} \int_D f(x) \cos \frac{2\pi nx}{D} dx \qquad (3.03)$$

where $n = 0, 1, 2, \ldots$.

The expression for the constant term, $A_0/2$, is simply the average value of $f(x)$ over the whole pattern-unit, of repeat D. This confirms and quantifies the observation made in §3.2 that $A_0/2$ is a constant amount by which the summation has to be raised from the x-axis to match the height of the profile of $f(x)$.

For $n \neq 0$, the interpretation of Eqn (3.03) is fascinating. As illustrated graphically in Fig. 3.02 (though normally it would be done by computation) for the function $f(x)$ that we are already familiar with, it means that if we wish to know whether a cosine of spatial frequency $1/D$ say, is present in the Fourier series representation of $f(x)$, we multiply the pattern-unit of $f(x)$ by a cosine of the frequency of interest, i.e. we multiply it by $\cos 2\pi x/D$, and then integrate over D. The non-zero value obtained by the integration (i.e. the shaded area under the product curve) shows that in this example the cosine

Fig. 3.02 — Graphical construction of $A_1 = \dfrac{2}{D} \displaystyle\int_D f(x) \cos(2\pi x/D)\, \mathrm{d}x$.

of frequency $1/D$ is indeed a constituent of $f(x)$. The area, after multiplication by $2/D$ as required by Eqn (3.03), gives its amplitude. The sign of the cosine is revealed by the sign of the value obtained — the nett area under the product curve is negative in this example. (We already know from Fig. 3.01 that the $A_{n=1}$ term has a phase of π.)

For our purposes there is an advantage in rewriting Eqn (3.02) in the form

$$f(x) = \sum_{n=-\infty}^{+\infty} \frac{A_n}{2} \cos \frac{2\pi nx}{D} \;. \tag{3.04}$$

The term for $n = 0$ is the same as before. For other values of n the coefficients are now in pairs, $\frac{1}{2}A_{+n}$ and $\frac{1}{2}A_{-n}$, which are equal $(\cos(-x) = \cos(+x))$ and have the same spatial frequencies. They therefore add to give the same harmonic amplitude as when only positive values of n were involved. It follows that Eqn (3.03) now gives the coefficients $\frac{1}{2}A_n$ in Eqn (3.04) for all n, viz.

$$\frac{A_n}{2} = \frac{1}{D} \int_D f(x) \cos \frac{2\pi nx}{D}\, \mathrm{d}x \;. \tag{3.05}$$

The reason for modifying the series to consist of pairs of harmonics of half the original amplitude, but for negative as well as positive values of n, is because these pairs are seen in the following section to have a physical counterpart in the pairs of diffraction maxima of order $\pm n$ given by a grating.

General formulae for $f(x)$ when this is not an even function are given in §3.5.

3.4 OPTICAL AND CRYSTAL DIFFRACTION GRATINGS: PHYSICAL INTERPRETATION OF FOURIER TERMS

3.4.1 Optical diffraction

The grating whose overall aperture function, or optical structure as we are also describing it, is represented by $f(x)$ in Fig. 3.03(a) has slits of width a and

Fig. 3.03 — Fourier analysis of a periodic rectangle-function.

grating repeat D. The (uniform) amplitude of illumination at the slits is h.

The Fourier analysis of $f(x)$ is given by substituting these details into Eqn (3.05) as follows:

$$\frac{A_n}{2} = \frac{1}{D}\left[\int_0^{a/2} h\cos\frac{2\pi nx}{D}\,\mathrm{d}x + \int_{D-a/2}^{D} h\cos\frac{2\pi nx}{D}\,\mathrm{d}x\right]$$

whence

$$\frac{A_n}{2} = \frac{ha}{D}\left[\frac{\sin\dfrac{\pi na}{D}}{\dfrac{\pi na}{D}}\right]. \tag{3.06}$$

Plotted in Fig. 3.03(b), they give the amplitudes and phases of the pairs ($\pm n$) of harmonics of repeat D/n (**spatial frequency** n/D) comprising $f(x)$. Graphically, and in Eqn (3.06), they are also immediately recognizable as the **sampled sinc function** that tells us the amplitudes and phases of the pairs of diffraction maxima, of order $\pm n$, given by a grating whose overall aperture function is $f(x)$ (cf. the intensity values in Eqn (2.17)). And the directions of those diffraction maxima are given by the spatial frequencies of the Fourier harmonics because

$$\frac{n}{D} = \frac{\sin\theta}{\lambda}\quad(=u)\ . \tag{2.10}$$

The above findings are applicable to practically any type of aperture function, though of course the envelope of the coefficients is not generally a sinc function.

We have the important general result then, that the *amplitude* of the nth order pair of diffraction maxima from a grating is a measure of the amplitude of the nth order pair of harmonics comprising its overall aperture function — what we have called its optical structure. The *direction* of those diffraction maxima gives, as $(\sin\theta)/\lambda$, the spatial frequency (n/D) of the harmonics (or the harmonic repeat, D/n).

For these reasons the plane of a diffraction pattern is referred to as the **Fourier plane** (or **Fourier space**), or alternatively as the **(spatial) frequency plane** (or **frequency domain**). Furthermore, as first noted in Chapter 2, the term **reciprocal space** can also be used, because of the reciprocal relationship between the scale of a diffracting system and the pattern it gives. Each of these interpretations has its particular relevance and usefulness, e.g. the 'frequency domain' is widely used in connection with optical data processing (§5.5).

3.4.2   Crystal diffraction
In §2.7 it was shown that the amplitude, $A_n(\theta)$, of a Bragg reflection of order n, at crystal lattice planes of spacing d, is given by

$$A_n(\theta) \;=\; \int_{x=0}^{x=d} \rho(x)\, e^{i(nx/d)2\pi}\, dx \qquad\qquad \{2.21\}$$

where θ is related to d by the Bragg equation.

Let us apply this to the crystal structure of caesium chloride, CsCl (Fig. 3.04). The electron density distribution through the structure in the direc-

Fig. 3.04 — The crystal structure of caesium chloride, CsCl.

tion of one of the cube edges can be represented by the curve $f(x)$ used in Fig. 3.01(a); it was deliberately chosen for this purpose. The low peaks in that pattern represent the electron cloud of the chloride ions and the high peaks that of the caesium ions. (There are equal numbers of each in the structure.)

If we ignore various angle-dependent (and other) factors, we can assume for the present purpose that the amplitude of X-ray scattering $\rho(x)$ in the above equation is equal to $f(x)$, and since this particular $f(x)$ is an even function we can rewrite the above equation as

$$A_n(\theta) = \int_{x=0}^{x=d} f(x) \cos\frac{2\pi nx}{d} \, dx \; . \tag{3.07}$$

Comparison with Eqn (3.05) shows that these values for the diffraction amplitudes are proportional to the amplitudes of the harmonic terms of the Fourier series analysis of the electron density distribution between the reflecting planes in the crystal structure.

The same relationship applies to the full 3-dimensional nature of crystal structures and it is readily extended to deal with examples that are not centrosymmetric. Thus we have the fundamental and outstandingly import-ant result that measurement of X-ray diffraction maxima — more conve-niently regarded as Bragg reflections — provides information about the terms in the Fourier series analysis of the crystal structure, just as with the 'structure' of an optical diffraction grating. An experimental snag is that because only the intensities of the diffracted X-rays can be measured the phases of the Fourier terms are unknown and a Fourier synthesis cannot be performed directly (§5.3.3).

The notes at the end of the previous section concerning the ways of describing the space in which a diffraction pattern is formed are equally applicable here. The concept of reciprocal space is particularly useful in X-ray crystallography.

3.5 FOURIER SERIES: GENERAL FORMULATIONS

When $f(x)$ is a general periodic function the α_n in Eqn (3.01) can have values from 0 to 2π. The following two alternative forms of the series are then often more useful.

3.5.1 The sine and cosine series
Starting with the cosine series in Eqn (3.01)

$$f(x) = \frac{A_0}{2} + \sum_{n=1}^{\infty} A_n \cos\left(\frac{2\pi nx}{D} - \alpha_n\right)$$

we can expand this as follows:

$$f(x) = \frac{A_0}{2} + \sum_{n=1}^{\infty} \left[A_n \cos \alpha_n \cdot \cos \frac{2\pi nx}{D} + A_n \sin \alpha_n \cdot \sin \frac{2\pi nx}{D} \right] ,$$

which in turn can be written as

$$f(x) = \frac{A_0}{2} + \sum_{n=1}^{\infty} a_n \cos \frac{2\pi nx}{D} + \sum_{n=1}^{\infty} b_n \sin \frac{2\pi nx}{D} \tag{3.08}$$

where $a_n = A_n \cos \alpha_n, \ b_n = A_n \sin \alpha_n$.

The original series of cosines with various phases has thus been replaced by a combination of cosines and sines (cf. $2 \cos(2\pi nx/D - \pi/3) = 1 \cos 2\pi nx/D + \sqrt{3} \sin 2\pi nx/D$).

The following expressions for the new coefficients in (3.08) are obtained in a similar way to that used for the original single series and we have:

$$\left. \begin{aligned} A_0 &= \frac{2}{D} \int_D f(x) \, dx \\[2ex] a_n &= \frac{2}{D} \int_D f(x) \cos \frac{2\pi nx}{D} \, dx \\[2ex] b_n &= \frac{2}{D} \int_D f(x) \sin \frac{2\pi nx}{D} \, dx \ . \end{aligned} \right\} \tag{3.09}$$

The reader can check that for an even function all $b_n = 0$ and that, similarly, for an odd function all $a_n = 0$.

3.5.2 Exponential notation

For $f(x)$ real we can start with Eqn (3.01) as we did at the beginning of the previous section but now we will rewrite it as

$$f(x) = \frac{A_0}{2} + \sum_{n=1}^{\infty} \frac{1}{2} A_n \, e^{i \left[\frac{2\pi nx}{D} - \alpha_n \right]} + \sum_{n=1}^{\infty} \frac{1}{2} A_n \, e^{-i \left[\frac{2\pi nx}{D} - \alpha_n \right]} .$$

The negative sign in front of i in the second summation may be combined with the n, making it a summation over negative values of n.

The equation can therefore be changed to

$$f(x) = \frac{A_0}{2} + \sum_{n=1}^{\infty} \frac{1}{2} A_n e^{i\left[\frac{2\pi nx}{D} - \alpha_n\right]} + \sum_{n=-1}^{-\infty} \frac{1}{2} A_n e^{i\left[\frac{2\pi nx}{D} + \alpha_n\right]}$$

which can be compressed to

$$f(x) = \sum_{n=-\infty}^{+\infty} C_n e^{\frac{i2\pi nx}{D}} . \tag{3.10}$$

The complex coefficients C_n contain the A_n and α_n as follows

$$C_{n>0} = \frac{1}{2} A_n e^{-i\alpha_n}$$

$$C_{n<0} = \frac{1}{2} A_n e^{+i\alpha_n} \tag{3.11}$$

$$C_{n=0} = \frac{1}{2} A_0 .$$

A consequence of this is that

$$C_{n<0} = C^*_{n>0}$$

and therefore

$$|C_{n<0}| = |C_{n>0}| \tag{3.12}$$

where * denotes the complex conjugate.

The coefficients are derived in an analogous way to that for the cosine series and we have the pair of equations

$$f(x) = \sum_{n=-\infty}^{+\infty} C_n e^{\frac{i2\pi nx}{D}} \tag{3.10}$$

$$C_n = \frac{1}{D} \int_D f(x) e^{\frac{-i2\pi nx}{D}} dx . \tag{3.13}$$

As an example, consider the exponential Fourier series for the function

$f(x) = \cos(2\pi 5x/D)$. Of course this expression could be substituted into Eqn (3.13) and the Fourier coefficients C_n obtained. However, the answer is obtainable by inspection because

$$\cos\frac{2\pi 5x}{D} = \frac{1}{2}\left[e^{\frac{i2\pi 5x}{D}} + e^{\frac{-i2\pi 5x}{D}}\right].$$

The exponential series for $f(x)$ consists of only these two terms, viz. terms for which $n = \pm 5$ with coefficients 1/2.

Similarly, if $f(x) = \sin(2\pi 5x/D)$ only terms $n = \pm 5$ are non-zero, and the coefficients are $\pm \frac{1}{2}i$ respectively.

Figure 3.05 shows these two examples in graphical form.

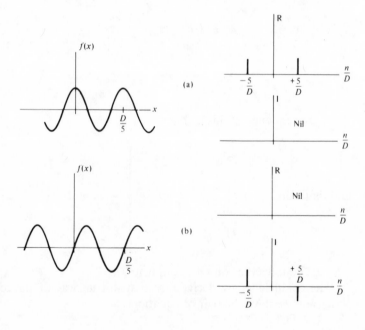

Fig. 3.05

Eqn (3.13) was derived from Eqn (3.01), i.e. for $f(x)$ real, and the presence of imaginary numbers only arises in the way that we have just seen, the whole expression being still real.

The above examples also illustrate the condition stated in Eqn (3.12). Thus for real functions that are even, the pairs of coefficients are real and of the same sign, whereas for real functions that are odd, the pairs are imaginary and of opposite sign.

For **complex functions** Eqn (3.12) does not apply. This is quickly illustrated by considering the function

$$f(x) = \sin\frac{2\pi 5x}{D} + \frac{i}{2}\cos\frac{2\pi 5x}{D}$$

for which the coefficients are $C_5 = -i/4$, $C_{-5} = 3i/4$.

As a final example of the exponential Fourier series the reader should derive the coefficients for the repeating rectangle function in Fig. 3.03(a). Inserting the details of that function into Eqn (3.13)

$$C_n = \frac{1}{D}\left\{\int_0^{a/2} h\,e^{\frac{-2\pi inx}{D}}\,dx + \int_{D-a/2}^{D} h\,e^{\frac{-2\pi inx}{D}}\,dx\right\}$$

$$= \frac{h}{D}\left\{\left[\frac{-D}{2\pi in}\,e^{\frac{-2\pi inx}{D}}\right]_0^{a/2} + \left[\frac{-D}{2\pi in}\,e^{\frac{-2\pi inx}{D}}\right]_{D-a/2}^{D}\right\}$$

$$= \frac{-h}{2\pi in}\left\{\left[e^{\frac{-\pi ina}{D}} - 1\right] + \left[e^{-2\pi in} - e^{\frac{-2\pi in}{D}[D-(a/2)]}\right]\right\}$$

$$= \frac{-h}{2\pi in}\left\{\left[e^{\frac{-\pi ina}{D}} - 1\right] + \left[1 - e^{\frac{\pi ina}{D}}\right]\right\}$$

$$= \frac{h}{\pi n}\sin\frac{\pi na}{D}$$

i.e.

$$C_n = \frac{ha}{D}\left[\frac{\sin\frac{\pi na}{D}}{\frac{\pi na}{D}}\right] \tag{3.14}$$

Here is the sampled sinc function as before (Eqn (3.06)). Since the function is real and even, the coefficients are real and equal in pairs ($\pm n$), as found in §3.3 using the cosine series.

3.5.3 Space and time

The Fourier series described in this chapter are readily extended to 2- and 3-dimensions. Also, when the patterns analysed are functions in time instead of space the Fourier coefficients refer to temporal frequencies instead of spatial frequencies. The corresponding equations in exponential notation are then

$$f(t) = \sum_{n=-\infty}^{+\infty} C_n e^{\frac{i2\pi nt}{T}} \tag{3.15}$$

$$C_n = \frac{1}{T} \int_T f(t) e^{\frac{-i2\pi nt}{T}} dt . \tag{3.16}$$

For example, if the repeating rectangle function we have used represents a stream of pulses of duration a and repeat time T, then the Fourier terms form the frequency spectrum required for the generation of the pulses.

In §4.6 the application of Fourier methods to a single time-pulse leads to an interpretation of the frequency spectrum of the wavetrains associated with light photons.

PROBLEMS

3.1 Construct each term in the following expression for $f(x)$, graphically, for one repeat of D:

$$f(x) = 3 - \cos(2\pi x/D) + 2\cos(4\pi x/D) .$$

Combine the three terms graphically to form $f(x)$. Then use the graph you have obtained, to derive the values of A_0, a_n and b_n in the Fourier series representation of $f(x)$ given by Eqn (3.08), for $n = 2$ and $n = 3$. To do this, use Eqn (3.09) *graphically*, performing the integrations by counting squares.

3.2 The periodic function

$$f(x) = \sin(2\pi 5x/D) + (i/2)\cos(2\pi 5x/D)$$

is used in §3.5 to illustrate how, with *complex* functions, the Fourier coefficients are not of equal magnitude. By expressing the sine and cosine as exponentials show that the coefficients are as given in the text, (a) by inspection and (b) analytically by substituting for $f(x)$ in Eqn (3.13).

4

Fourier transforms, convolution and correlation

4.1 INTRODUCTION

In Chapter 3 the harmonic terms comprising the Fourier analysis of the optical structure of an object in the form of a multiple-aperture grating have been identified with details of the diffraction pattern given by the grating. There is an analogous relationship between a single aperture and its diffraction pattern, which we need to explore. It is the **Fourier transform**, involving — as one would expect for a non-repeating pattern — an integral rather than a series. This is the subject of §4.2.

The Fourier transform also has other important roles in physical optics; indeed it would be difficult to exaggerate its importance generally in physics. This chapter is concerned with the further insight the Fourier transform provides into the relationship between the diffraction pattern given by a multiple-aperture diffracting system such as a grating or crystal, and the (overall) aperture function of that diffracting system. The basic ideas of this are introduced in §4.3–§4.5 for use in various ways in Chapter 5 in connection with image formation and image processing. In §4.3 we consider the grating diffraction pattern, and in §4.4 the aperture function of the grating. The latter is considered in terms of convolution — another concept and mathematical procedure widely used in physics. In §4.5 the two sides of the relationship, the aperture function of the grating and the diffraction pattern it gives, are brought together and seen as an example of the convolution theorem.

In §4.6 we consider something quite different, the role of the Fourier transform in relating the length of a wavetrain of light to the spectral composition of the light. This is of importance when we turn our attention in Chapter 7 to the analysis of radiation sources by interferometry.

Finally in §4.7, we introduce correlation, an operation of considerable importance in both Chapters 5 and 7.

4.2 THE FOURIER TRANSFORM AND SINGLE-SLIT DIFFRACTION

First let us recall the repeating rectangle function (Fig. 4.01(a)). With the

Fig. 4.01 — Fourier analysis of a periodic rectangle-function.

chosen origin this is an even function, and in §3.4.1 it was noted that if an even function is represented by the Fourier series

$$f(x) = \sum_{n=-\infty}^{+\infty} \frac{A_n}{2} \cos \frac{2\pi nx}{D} \qquad \{3.04\}\ (4.01)$$

then the coefficients,

$$\frac{A_n}{2} = \frac{1}{D} \int_D f(x) \cos \frac{2\pi nx}{D} \,dx \qquad \{3.05\}\ (4.02)$$

are given by

$$\frac{A_n}{2} = \frac{ha}{D}\left[\frac{\sin\dfrac{\pi na}{D}}{\dfrac{\pi na}{D}}\right]. \qquad \text{\{3.06\} (4.03)}$$

This last expression contains what we have called a sampled sinc function and it is shown again in Fig. 4.01(a). Furthermore, if $f(x)$ represents the aperture function of a whole diffraction grating, of repeat D, we have identified (in §3.4.1) these A_n with the amplitude of light from each slit in just the directions ($\pm\,\theta$) of the grating maxima (order $\pm\,n$). Thus the term in [ ] brackets in Eqn (4.03) is the single-slit pattern, $(\sin \pi ua)/(\pi ua)$, in which u is restricted to specific values given by

$$u = \frac{\sin\theta}{\lambda} = \frac{n}{D} \qquad \text{\{2.10\} (4.04)}$$

$D \sin\theta = n\lambda$ being the condition for a grating maximum (§2.5).

We should recall, too, that we also have an interpretation of the spatial frequencies, n/D, of the harmonics in the Fourier analysis (Eqn (4.03)) of the grating aperture function. They tell us, via Eqn (4.04), the directions ($\pm\,\theta$) of the principal diffraction maxima given by the grating.

Now consider what happens if D is increased, as illustrated in Fig. 4.01(b). The spatial frequencies, and therefore the directions of the diffracted beams, become closer together. This suggests that for the limiting case of a non-periodic function, all frequencies are allowed. That is certainly consistent with our knowledge of diffraction: with a single aperture the diffraction pattern is a continuous function of u.

When D becomes infinite in Fig. 4.01(b) we are left with the single rectangle centred at $x = 0$. In Eqns (4.01) and (4.02) n and D lose their significance and we have the integrals

$$f(x) = \int_{-\infty}^{+\infty} F(u)\cos 2\pi ux \, du \qquad (4.05)$$

$$F(u) = \int_{-\infty}^{+\infty} f(x)\cos 2\pi ux \, dx \ . \qquad (4.06)$$

$f(x)$ and $F(u)$ are now said to be the **Fourier transform** of each other, in the present instance the **cosine Fourier transform** of each other because $f(x)$ is a real, even function (the general case for a real function is dealt with below, in (i)). Eqn (4.05) replaces the summation in (4.01), and Eqn (4.06) corresponds to (4.02). For the single rectangle that we are for the moment considering, Eqn (4.03) becomes

$$F(u) = ha\left(\frac{\sin \pi ua}{\pi ua}\right) \qquad (4.07)$$

in agreement with the phasor derivation of the diffraction pattern for a single slit (cf. Eqn (2.05) where $R_1(0)$ was the total amplitude of illumination at the slit, corresponding here to ha). Thus we can say that the single-slit diffraction pattern is the Fourier transform of the aperture function of the slit. This is true for any aperture function.

● ● ● ● ●

(i)   *Fourier transform of real $f(x)$: general case, using exponential notation*

When $f(x)$ is not necessarily even (or odd) we could, just as in §3.5.1, consider the general case in terms of sine and cosine integrals. However, exponential notation is generally more convenient, and it also has the advantage of being applicable to complex as well as real functions ((ii) below). By referring to Eqns (3.10) and (3.13) we can immediately write

$$f(x) = \int_{-\infty}^{+\infty} F(u)\, e^{2\pi iux}\, du \qquad (4.08)$$

$$F(u) = \int_{-\infty}^{+\infty} f(x)\, e^{-2\pi iux}\, dx \ . \qquad (4.09)$$

(It is immaterial which of these equations contains the negative sign in the exponential.)

At the end of this section we consider the application of this exponential formulation to the simple case of the rectangle function, and see that it gives the same result as we have already obtained.

The product of x and u in these equations, and in (4.05) and (4.06), is dimensionless: x and u are **conjugate parameters**. As noted already, $f(x)$ and $F(u)$ are each described as the **Fourier transform** (or 'integral') of the other. They constitute what is called a **Fourier transform pair**. (Conventionally, a distinction is made between 'transform' and 'inverse transform'. This arises from the negative sign in one exponential.)

(ii)   *Fourier transform of complex $f(x)$*

To see how Eqns (4.08) and (4.09) are used for complex functions, let us assume that we have $f(x)$ as a complex function and that we wish to calculate $F(u)$. We can express $f(x)$ as the sum of its real and imaginary parts as follows:

$$f(x) = \upsilon(x) + iw(x) \ . \tag{4.10}$$

Using Euler's relation ($e^{i\theta} = \cos\theta + i\sin\theta$), Eqn (4.09) then becomes

$$F(u) = \int_{-\infty}^{+\infty} \upsilon(x)\cos(2\pi ux)\,dx + \int_{-\infty}^{+\infty} w(x)\sin(2\pi ux)\,dx$$

$$+ i\left[\int_{-\infty}^{+\infty} w(x)\cos(2\pi ux)\,dx - \int_{-\infty}^{+\infty} \upsilon(x)\sin(2\pi ux)\,dx\right] \ . \tag{4.11}$$

Each integrand is real and $F(u)$ can be calculated in the usual way by combining the first two with the second two in the complex plane (Argand diagram).

(iii)   *Exponential form reduces to simple forms when $f(x)$ is real*

Finally, we will note how the exponential form reduces to simple forms when $f(x)$ is real, and to Eqn (4.05) in particular, when it is also even. To do this we put

$$\left.\begin{aligned} F(u) &= A(u) + iB(u) \\[4pt] \text{where} \quad A(u) &= F(u)\cos\alpha(u) \\ B(u) &= F(u)\sin\alpha(u) \\[8pt] \text{and} \quad \alpha(u) &= \tan^{-1}\frac{B(u)}{A(u)} \ . \end{aligned}\right\} \tag{4.12}$$

Then we can write Eqn (4.08) as

$$f(x) = \int_{-\infty}^{+\infty} (A(u) + iB(u))[\cos(2\pi ux) + i\sin(2\pi ux)]\,du$$

$$= \int_{-\infty}^{+\infty} [(A(u)\cos(2\pi ux) - B(u)\sin(2\pi ux)]\,du$$

$$+ i\int_{-\infty}^{+\infty} [(B(u)\cos(2\pi ux) + A(u)\sin(2\pi ux)]\,du \ . \tag{4.13}$$

If $f(x)$ is real then the imaginary term must be zero. This requires that

$$B(-u) = -B(+u)$$

$$A(-u) = A(+u)$$

which means that

$$F(u) = A(u) + iB(u) , \qquad u > 0$$

but $\qquad F(u) = A(u) - iB(u) , \qquad u < 0$

i.e. $\qquad\qquad F(u) = F^*(-u)$

$\left.\begin{array}{l} \\ \end{array}\right\}$ (4.14)

and therefore $\qquad |F(u)| = |F(-u)|$.

Thus for $f(x)$ real we have simply the first term in Eqn (4.13) i.e.

$$f(x) = \int_{-\infty}^{+\infty} A(u)\cos(2\pi ux)\,du - \int_{-\infty}^{+\infty} B(u)\sin(2\pi ux)\,du \quad (4.15)$$

or, with the help of Eqns (4.12), we can express this as

$$f(x) = \int_{-\infty}^{+\infty} F(u)\cos[2\pi ux + \alpha(u)]\,du$$

$$= 2\int_{0}^{\infty} F(u)\cos[2\pi ux + \alpha(u)]\,du$$

$\left.\begin{array}{l} \\ \\ \end{array}\right\}$ (4.16)

where $F(u)$ is restricted by Eqn (4.14). The first form of (4.16) reduces to Eqn (4.05) when $f(x)$ is even.

As with Fourier series, transforms are equally applicable to 2- and 3-dimensions. They are also used with time and frequency as the conjugate pair (§4.6).

•   •   •   •   •

To return to the case of the single rectangle function (Fig. 4.02(a)), the

Fig. 4.02 — Fourier transform pairs, $f(x)$ and $F(x)$.

exponential notation of Eqn (4.09) gives the expected sinc function quite simply. We have

$$F(u) = \int_{-a/2}^{+a/2} h\, e^{-2\pi i u x}\, dx \tag{4.17}$$

$$= \frac{h}{-i2\pi u}\left[e^{-i2\pi u x} \right]_{-(a/2)}^{+(a/2)}$$

$$= \frac{h}{-i2\pi u}\left(e^{-i\pi u a} - e^{+i\pi u a} \right)$$

$$= \frac{h}{\pi u}\left(\frac{e^{i\pi u a} - e^{-i\pi u a}}{2i} \right)$$

$$= \frac{h}{\pi u} \sin \pi u a$$

$$= ha\left(\frac{\sin \pi u a}{\pi u a} \right) \tag{4.18}$$

in agreement with the previous derivation (Eqn. 4.07).

Note that the Fourier transform $F(u)$ (Eqn (4.17)) can be identified directly as the amplitude of diffracted light, in a direction specified by u, from a uniformly illuminated slit of width a. This is evident from Fig. 4.03 where we see that the contribution from an element dx of wavefront at distance x from the origin (0) has an amplitude $h\, dx$ and a phase of $(2\pi/\lambda)x \sin\theta = 2\pi u x$, so that for the whole slit the nett amplitude is

$$\int_{-a/2}^{+a/2} h\, e^{2\pi i u x}\, dx \ .$$

Fig. 4.03 — Single-slit Fraunhofer diffraction.

In connection with the Fraunhofer diffraction pattern given by a single aperture being the Fourier transform of its aperture function, note that the 'symmetry' in the relationship between $f(x)$ and $F(u)$ (seen so strikingly in Eqns (4.08) and (4.09)) is such that if $f(x)$ were an aperture function in the form of the sinc function, then its diffraction pattern, $F(u)$, would be the rectangle function: this exchange of functions is illustrated in Fig. 4.02 for the sinc and rectangle functions but it is of general applicability.

We need also to note that the earlier findings (§2.4, §2.5) concerning the sampling relationship between the diffraction pattern given by a grating and by a single aperture may now be extended to describe a grating pattern as a sampling of the Fourier transform of the single aperture function.

And as both the Fourier transform of a single aperture and the Fourier series of a grating are spatially specified by u, continuous in one and discrete in the other, both single-aperture and grating diffraction patterns may be described as existing in **Fourier space**, or **frequency space** (or **domain**), as stated in §3.4.1 in connection with the diffraction grating. Furthermore, since a grating is inevitably of finite extent and its (overall) aperture function can be regarded as a single, non-repeating unit (§3.2), in that sense even a grating diffraction pattern can be described as the Fourier transform of the grating aperture function.

The interpretation of the Fraunhofer diffraction pattern given by *any* aperture function as the Fourier transform of that aperture function is a most valuable concept and we shall use it freely. (It is interesting to realize that it leads to the description of a lens as a **Fourier transformer**: we take lenses so much for granted that it is easy to underestimate this remarkable property of what is after all a lump of glass.)

4.3   THE GRATING PATTERN AS A PRODUCT OF TRANSFORMS

In the following expression for the diffraction pattern given by an N-slit grating

$$R_N = R_1(0) \left(\frac{\sin \pi u a}{\pi u a} \right) \left(\frac{\sin N \pi u D}{\sin \pi u D} \right) \qquad \{2.15\} \ (4.19)$$

the '**single-slit**' term, $(\sin \pi u a)/\pi u a$, has been seen in the previous section to be the Fourier transform of the aperture function of the slit.

Turning our attention now to the '**grating term**' in this expression we shall find that it also is a Fourier transform, this time of the grating 'lattice' on which the individual slits are distributed. The result, that the diffraction pattern represented by Eqn (4.19) is therefore the product of two transforms, is of general validity for other forms of aperture function and for gratings of 1, 2, and 3 dimensions. (The description, in the previous section, of the grating diffraction pattern as the Fourier transform of the grating aperture function in no way conflicts with this: on the contrary, it leads to the convolution theorem (§4.5).)

To establish the Fourier transform nature of the grating term the regular lattice on which the grating is built is specified by an array of 'markers' at each one of which is located an identical single aperture — here a slit. For such a marker we use the Dirac **δ-function**, a mathematical device (not a true function in the mathematical sense) defined as the limiting form of a rectangle function (Fig. 4.02(a)) when, whilst maintaining constant area (usually chosen as unity), its width tends to zero while the height tends to infinity. The δ-function thus has a value of zero except at one specific point where it is infinite. In some contexts it is described as a (unit) **impulse function**. (No known ordinary function behaves like this and it is described as a generalized function: it is normally specified by the integral properties attached to it.)

First consider a single δ-function located at the origin of x in Fig. 4.04(a): it is denoted as the function $f(x) = \delta(x)$. The Fourier transform, $F(u)$, of this δ-function is easily deduced by referring to Fig. 4.02(a). As the width, a, of the rectangle function becomes smaller we can see that the central peak of its Fourier transform (the sinc function) becomes broader. In the limit, when $a = 0$ the first zero in the $F(u)$ curve is at infinity. So we have the result (Fig. 4.04(a)) that the Fourier transform of a δ-function at the origin is ha for all u, or unity if we adopt the convention of putting $ha = 1$.

This is what one would expect for the diffraction pattern from an 'ideally narrow' slit (a slit of infinitesimal width), as there would be no path

Fig. 4.04 — δ-functions and their Fourier transforms.
(a) δ-function at the origin.
(b) δ-functions at $x = \pm D/2$.

differences like those causing interference in the light diffracted from a slit of finite width. In practice, such a slit would transmit an infinitesimal amount of light and it is in that sense that it is described above as a marker.

To build up an array of δ-functions to represent a grating it is necessary to specify δ-functions at other values of x. A δ-function moved from the origin to some position x_1 is expressed as

$$f(x) = \delta(x - x_1)$$

and its Fourier transform is

$$F(u) = \int_{-\infty}^{+\infty} \delta(x - x_1) \, e^{2\pi iux} \, dx \ .$$

Put $x - x_1 = x'$, then $dx = dx'$ and we have

$$F(u) = \int_{-\infty}^{+\infty} \delta(x') \, e^{2\pi iu(x' + x_1)} \, dx'$$

$$= e^{2\pi iux_1} \int_{-\infty}^{+\infty} \delta(x') \, e^{2\pi iux'} \, dx'$$

$$= e^{2\pi iux_1} \tag{4.20}$$

i.e. unit modulus with phase depending on x_1. This means that in terms of the δ-function representing an ideally narrow slit, and $F(u)$ being its diffraction pattern in the frequency domain, we have the result that translation of the slit in the spatial domain introduces a linear phase shift in the frequency domain (diffraction pattern). We meet another example of this '**shift theorem**' below.

A multiple array in one dimension is similarly written as

$$f(x) = \sum_N \delta(x - x_N) \tag{4.21}$$

and its Fourier transform is

$$F(u) = \sum_N e^{2\pi i u x_N} . \tag{4.22}$$

As an example, consider two δ-functions at $x = \pm D/2$. The transform is then

$$F(u) = e^{\pi i u D} + e^{-\pi i u D}$$

$$= 2 \cos \pi u D .$$

This particular Fourier pair, $f(x)$ and $F(u)$, is illustrated in Fig. 4.04(b). $F(u)$ is readily interpretable as the diffraction pattern for two ideally narrow slits whose positions are specified by the δ-function markers defined by $f(x)$: it is the grating term in Eqn (4.19) for $N = 2$. Just as for the single δ-function, we can show that if this pair of δ-functions were moved along the x-axis there would be an overall alteration of phase throughout the diffraction pattern (cf. the **shift theorem** referred to above). Let us move them by $D/2$, to $x = 0$ and $x = D$ respectively. Then

$$F(u) = \sum_N e^{2\pi i u x_N} = e^0 + e^{2\pi i u D}$$

$$= e^{\pi i u D}(e^{-\pi i u D} + e^{\pi i u D})$$

i.e.

$$F(u) = 2 e^{2\pi i u D/2} \cos \pi u D ,$$

where the exponential term carries the phase change corresponding to the δ-functions being moved through $D/2$. However, as it is the intensity distribution that is normally observed and measured, the pattern would *appear* unchanged: multiplying the above amplitude by its complex conjugate gives $4\cos^2\pi uD$ for the intensity, as though the apertures had not been moved. But in fundamental terms, phase is just as important as amplitude in a diffraction pattern, and this cannot be stressed too much because with phase it is a matter of 'out of sight out of mind' — except in special circumstances such as holography.

In the same way, the Fourier transform of a linear array of N equally spaced δ-functions is the **grating term** for N equally spaced slits (Appendix D).

We thus have the result that the two terms in Eqn (4.19), the 'single-slit term' and the 'grating term', are respectively the Fourier transform of the single-aperture function and the transform of the array of δ-functions defining the distribution of the apertures. That the diffraction pattern is the product of these two transforms is an example of the **array theorem**, a theorem that owed its name originally to the analogous situation that arises in the theory of multiple radio antennae, where δ-functions are similarly used.

4.4   CONVOLUTION

4.4.1   Introduction

In the previous section the aperture function of an entire grating was described as a distribution of its constituent single-aperture function in accordance with an array of δ-functions defining the lattice on which the grating is based. This distribution of one entity in a manner specified by another is an example of **'convolution'**. It is a process that occurs in many forms and in numerous contexts. A Fourier aspect of convolution, embodied in the convolution theorem (§4.5), is of wide-ranging importance and usefulness — nowhere more so than in the areas of image formation and processing that are our concern here.

There are special features of convolution that are not particularly clearly illustrated by the example of the grating because of the discontinuous nature of δ-functions. As an introduction we will consider a more common type, the ubiquitous one in which convolution occurs when a continuous 'input' signal to a system is processed to give an 'output'.

Imagine that the function $f(x)$ in Fig. 4.05(a) represents a 1-dimensional input signal to some device. It could, for example, be the way the voltage to an amplifier varies with time: instead of $f(x)$ we would label it $f(t)$. Or it could be the spatial distribution of light intensity in a certain direction, x, across a screen. In the first of these the amplifier is probably required to increase (amplify) every ordinate in (a) by some factor, to form the output. In the second example, we may imagine that a photometer is faithfully to scan and measure the intensity distribution $f(x)$ and display it in some way. In both cases, however, the processing of the 'input' is usually less than perfect in practice. Every ordinate of the input becomes 'smeared' rather than cleanly

Fig. 4.05 — Convolution of $f(x)$ and $g(x)$.

reproduced as another ordinate in the output. Thus a δ-function input at $x = 0$ yields an output, $g(x)$ say, instead of a δ-function.

We shall assume that the 'shape' of the response of the system, including this unwanted smearing, is **invariant**, i.e. it is the same at every value of x. It is represented by the triangle in Fig. 4.05(b), shown at just two values of x, where it is $g(x)$ and $g(x - x_1)$ respectively, for unit strength of input. Consider what happens to the input at x_1 in (a). The input ordinate there is $f(x_1)$ and it is smeared by the response of the system to become the curve $f(x_1)g(x - x_1)$ shown by the full line in (c); thus $f(x_1)$ is the 'weighting' of the unit response $g(x - x_1)$. Only the ordinate at x_1 in this smearing actually contributes to the output at x_1. This output ordinate is not drawn in the figure because it is necessary to see that there are other contributions there, ones that originate from the smearing of other ordinates in $f(x)$. One such contribution arises from the smearing of the ordinate at x' in (a), shown by the broken curve in (c). The contribution this makes to the output at x_1 is the ordinate $f(x')g(x_1 - x')$.

In this way the total output at x_1, shown as h in (d), is the sum of the effects at x_1 of the smearing of all ordinates in $f(x)$; in the example shown only ordinates relatively near to x_1 will have an effect because $g(x)$ is fairly narrow. The total output at x_1 is given by the integration

$$h(x_1) \; = \; \int_{x' = -\infty}^{x' = +\infty} f(x')g(x_1 - x')\,\mathrm{d}x' \; . \qquad (4.23)$$

The same applies for any value of x_1 and we can therefore express the output $h(x)$ as

$$h(x) \; = \; \int_{x' = -\infty}^{x' = +\infty} f(x')g(x - x')\,\mathrm{d}x' \; . \qquad (4.24)$$

This integral is described as the **convolution** of $f(x)$ and $g(x)$, both of which can be complex-valued functions. For short, convolution is denoted as

$$h(x) \; = \; f(x) \circledast g(x)$$

or simply　　$h(x) \; = \; f \circledast g \; .$

Other names for convolution, though this is the most commonly used, include **folding product** (cf. the German *Faltung*), **composition product**, and **super-position integral**. In the appropriate context $g(x)$ would be referred to as a **smoothing** (or **blurring**) **function**, **line-spread function** (LSF) or, in two dimensions, the **point-spread function** (PSF).

The example we have been considering illustrates how *convolution is the distribution of one function in accordance with a law specified by another function*. It involves multiplying each ordinate of a function by the whole of another function and summing the results. Note that it is assumed that, as often occurs in practice, we are dealing with a **linear system** — by which we mean that the output of the system is a linear superposition of the outputs resulting from each individual component in the input. Also note that, as mentioned earlier in this section, we have assumed '**invariance**' of the function $g(x)$. We return to these matters in §5.1.

So far, this has been essentially a physical picture of what happens. To calculate the output at any particular value of x, such as x_1, it is necessary to perform the integration in Eqn (4.23). This is represented graphically in Fig. 4.05(e) where x' is the dummy variable of the integration. $f(x')$ is multiplied by $g(x_1 - x')$, and the area (shaded) under the product curve is the value, $h(x_1)$, of the output at x_1. Note that before multiplication and integration, $g(x)$ is both displaced and reflected (cf. the term 'folding product').

Convolution is commutative, i.e.

$$f \circledast g = g \circledast f .$$

It is also associative

$$f \circledast (g \circledast h) = (f \circledast g) \circledast h ,$$

and distributive over addition

$$f \circledast (g + h) = (f \circledast g) + (f \circledast h) .$$

4.4.2 The grating as a convolution

A set of δ-functions specifying a one-dimensional grating lattice is represented by $f(x)$ in Fig. 4.06, and the aperture function of one slit, by $g(x)$. In the above section we have seen that the aperture function of the whole grating is the convolution, $h(x)$, of $f(x)$ with $g(x)$.

This example, like those in the previous section, illustrates how convolution may be visualized as the nett effect of distributing one function in accordance with a law specified by another function. In the previous example (Fig. 4.05) the distribution caused overlapping of ordinates to some extent, whereas with the grating example it does not. (A quite different example of convolution occurs in Chapter 5 in connection with image formation.)

Fig. 4.06 — A grating-aperture function, $h(x)$, as a convolution of a δ-function array, $f(x)$, with the single-aperture function, $g(x)$.

4.5   THE CONVOLUTION THEOREM AND DIFFRACTION

We can now bring together some of the results of the two previous sections and obtain a result that is of considerable importance.

In §4.4.2 it was seen that the aperture function of a whole grating can be described as the convolution of one single aperture function with the array of δ-functions that specifies the distribution of that aperture in the grating.

In §4.3 we saw that the grating diffraction pattern is given by the product of the Fourier transform of the single-aperture function and the transform of the array of δ-functions defining the grating.

Since the grating diffraction pattern can be interpreted as the Fourier transform of its overall aperture function (§4.2), we can now say that the Fourier transform of the convolution of the single aperture function with the δ-function array is equal to the product of the individual transforms. This is an example of the **convolution theorem** which states that the Fourier transform of the convolution of two functions is equal to the product of their individual transforms.

In the present context this leads to the very important result (Fig. 4.07) that convolution in object space (real space) corresponds to multiplication in diffraction space (i.e. Fourier, or reciprocal, space: also frequency space). This is not only of interpretative value but it also provides a powerful tool in the computational side of image and signal processing (§5.5 and Appendix I).

As the convolution theorem has been arrived at here in a rather special way, the reader may like to have a direct derivation.

We have used $h(x)$ to denote the convolution of two functions $f(x)$ and

Fig. 4.07 — An example of the convolution theorem. Convolution in object space corresponds to multiplication in diffraction space.

$g(x)$. Their individual Fourier transforms are conventionally written as $H(u)$, $F(u)$ and $G(u)$ respectively, where x and u are the usual conjugates.

The **convolution theorem** can therefore be expressed by saying that if

$$h(x) = f(x) \circledast g(x)$$

$$\text{then} \quad H(u) = F(u)G(u) \, . \qquad\qquad (4.25)$$

To show this, we start by using the definition of a Fourier transform to write

$$H(u) = \int_{-\infty}^{+\infty} h(x) \, e^{2\pi i u x} \, dx \, . \qquad\qquad (4.26)$$

Substituting for $h(x)$ as the convolution of $f(x)$ and $g(x)$, using Eqn (4.24), gives

$$H(u) = \int e^{2\pi i u x} \left[\int f(x')g(x-x')\,dx' \right] dx \ . \tag{4.27}$$

Changing the order of integration this becomes

$$H(u) = \int f(x') \left[\int e^{2\pi i u x} g(x-x')\,dx \right] dx' \ . \tag{4.28}$$

In the inner integral, in which x' is constant, we make the change of variable $x - x' = X$, then $dx = dX$ and we have

$$H(u) = \int f(x') \left[\int e^{2\pi i u(X+x')} g(X)\,dX \right] dx' \ . \tag{4.29}$$

The factor $e^{2\pi i u x'}$ is a constant so far as the inner integral is concerned; and the rest of this inner integral, not containing x', is a constant in the outer integration with respect to x'. Therefore the whole breaks up into the product of two separate integrals, giving

$$H(u) = \int_{-\infty}^{+\infty} f(x')\, e^{2\pi i u x'}\,dx' \int_{-\infty}^{+\infty} g(X)\, e^{2\pi i u X}\,dX \tag{4.30}$$

i.e. $$H(u) = T[f(x)] \times T[g(x)] \tag{4.31}$$

or $$H(u) = F(u)G(u) \tag{4.32}$$

where T denotes Fourier transformation.

4.6 FOURIER TRANSFORMS AND LIGHT WAVES

The comments made in §1.2 about the wave nature of light and the discontinuous nature of its production can now be consolidated. To do this, the Fourier transform is used in just the same way as in studying the relationship between an aperture function and the distribution of its constituent spatial frequencies. Instead of what can be regarded as a 'space pulse' we now have a pulse in time associated with a distribution (i.e. spectrum) of temporal frequencies. The conjugates x and u are replaced by time, t, and frequency, v. The Fourier transform equations can be written down directly by changing the symbols in Eqns (4.08) and (4.09) as follows

$$f(t) = \int_{-\infty}^{+\infty} F(v)\, e^{2\pi i v t}\, dv \tag{4.33}$$

$$F(v) = \int_{-\infty}^{+\infty} f(t)\, e^{-2\pi i v t}\, dt \ . \tag{4.34}$$

To illustrate the analogy between this new scenario and diffraction, Fig. 4.08(a) shows a rectangle function and its transform, now labelled

Fig. 4.08 — Time-frequency Fourier transforms.

according to the new variables. However, as we already know, a major component of this rectangle function is a non-oscillatory (i.e. zero frequency) constant amplitude — due to the function being entirely positive.

An example more appropriate to a consideration of light waves is the transform pair shown in Fig. 4.08(b). This shows a sinusoidal wave of frequency v_1 truncated to form a wavetrain of finite duration and length, and how it is then composed of an amplitude-frequency distribution spread about v_1 such that the summation gives a **wave group** (or **wave packet**) which represents the profile within the wavetrain but becomes out of step to give a zero nett amplitude on either side of it (Appendix G). When the wavetrain is long the frequency spread is narrow, and vice versa: there is the same reciprocity as with the spatial Fourier transform pair. Strictly monochromatic light would involve wavetrains of infinite length and as this is clearly not physically realizable, especially with light emitted discontinuously as **photons** by atoms, all spectral lines have a finite width. If, in Fig. 4.08(b), the 'width' of the frequency distribution is taken as essentially within $v_1 \pm \delta v$ we have

$$\delta v . \delta t \approx 1 . \tag{4.35}$$

This measure of width, the **bandwidth**, is used because most of the energy is associated with the central peak, as can be seen if the amplitude is squared to give the energy distribution (the **power spectrum**).

Since with **thermal** (i.e. non-laser) light sources the wavetrains emitted are not of identical duration, and the frequencies are modified by thermal motions, field effects, etc., it is necessary to think of an average duration time, Δt, and a spectral distribution somewhat different from that considered above. However, the reciprocity between pulse duration and frequency spread is fundamental and if the frequency bandwidth is Δv we have the so-called **Bandwidth Theorem**

$$\Delta v . \Delta t \approx 1 . \tag{4.36}$$

Δt in this equation is the **coherence time** of the light. The corresponding **coherence length**, l_c, is given by

$$l_c = c . \Delta t \approx c/\Delta v . \tag{4.37}$$

Using $c = \lambda v$ this gives

$$|l_c| \approx \lambda^2/\Delta\lambda . \tag{4.38}$$

Light sources such as discharge tubes give spectral linewidths of the order of 0.1 nm so that, taking $\lambda \approx 550$ nm, $l_c \approx 5500$ wavelengths. Although this is a large number of wavelengths it is physically a short length (about 3 mm) compared with the dimensions and displacements used in interferometry. And in Chapter 7 we indeed see how Fourier methods in spectroscopy are based on measurements of temporal coherence by interferometry.

In contrast to the above example, the linewidth of **laser light** is so narrow that l_c can be tens of kilometres, whereas the '**white light**' from a tungsten filament lamp has such a wide frequency range (Fig. 4.08(c)) that it consists of pulses of extremely short duration with no one frequency predominating ($-$ used by some authors as a definition of a **pulse**).

4.7 CORRELATION

Correlation is one of the most ubiquitous of all signal (optical and other) and data treatment tools in use today; and we see examples of its applications in Chapters 5 and 7. In its various manifestations it is essentially a method for assessing and specifying mutual relationships where these take the form of similarities or coincidences. An example is the 'correlation' between the phases of the Moon with the tides on the Earth: plotted as a function of time the two periodicities 'correlate'.

The process of testing for a correlation is basically one of comparing patterns; but it is not as straightforward as it may seem at first. An example will reveal the problem that arises. In Fig. 4.09(a) a pattern in the form of a

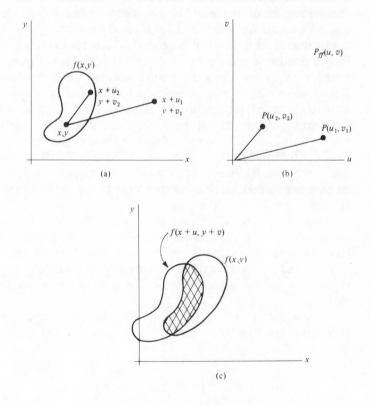

Fig. 4.09 — Autocorrelation.

'motif' is located in the x, y plane. Let it be described by a function $f(x, y)$, where f is some chosen property of the motif, e.g. its colour, composition, reflectance, transmittance, etc.; for simplicity we will assume that f is non-zero everywhere inside the motif and zero outside. Now this function $f(x, y)$ is no use whatever for describing the motif independently of the position of the motif in the x, y plane. Yet in many instances that is precisely what one needs to do. For example, in optical character recognition it is necessary to be able to identify the letter 'A' say, regardless of its position in a page of print. To do this a way of specifying the letter 'A' is needed that is 'invariant under translation'. Consider a function, $P_{ff}(u, \upsilon)$, that is the product of the value of f at x, y multiplied by its value at $x + u$, $y + \upsilon$, the product being summed for all x, y. This is expressed as

$$P_{ff}(u, \upsilon) = \int\int\limits_{-\infty}^{+\infty} f(x, y) f(x + u, y + \upsilon) \, dx \, dy \qquad (4.39)$$

Two examples of different values of u, υ will illustrate how the function P enables the motif in Fig. 4.09(a) to be described independently of x, y. Consider u_1, υ_1. The product of f at x, y and f at $x + u_1, y + \upsilon_1$, is zero for all x, y (just one pair is shown in the figure). This is because this particular u_1, υ_1 is such that for any x, y within the motif, where f is non-zero, the value at $x + u_1, y + \upsilon_1$, is zero since that point is always outside the motif. And if x, y is such that $x + u_1, y + \upsilon_1$ is inside the motif then x, y will be outside. The value of P at u_1, υ_1, plotted in (b), is therefore zero. For $P(u_2, \upsilon_2)$ the result is different because for at least some x, y the value of f at x, y and the value at $x + u_2, y + \upsilon_2$ are both non-zero (again only one such pair is shown in the figure). The value of $P(u_2, \upsilon_2)$ plotted in (b) therefore has some particular, non-zero value. P is thus a vector map that is unique to the motif, though not looking like it, and which is independent of the position of the motif in the x, y plane.

Note that Eqn (4.39) also gives us another way of visualizing the calculation of $P_{ff}(u, \upsilon)$. The equation shows that the value of P for any chosen u, υ is obtained by shifting the function $f(x, y)$ with respect to itself by $-u, -\upsilon$ and determining the area of overlap (Fig. 4.09(c). This function P is referred to as the **autocorrelation function** and is generally written as $P_{11}(u, \upsilon)$.

To compare two different motifs, $f(x)$ and $g(x)$, we can write a **cross-correlation function**

$$P_{12}(u, \upsilon) = \int\int\limits_{-\infty}^{+\infty} f(x, y) g(x + u, y + \upsilon) \, dx \, dy \qquad (4.40)$$

where P now refers to products between f and g.

Both correlation functions can be expressed in terms suitable for one, two or three dimensions and for time. To explore them in a little more detail we will revert to using one dimension:

$$P_{12}(u) = \int_{-\infty}^{+\infty} f(x)g(x+u)\,\mathrm{d}x \qquad (4.41)$$

and similarly with P_{11} for autocorrelation.

In order that u can describe a shift in the positive x direction the above equation can be rewritten as follows:

$$P'_{12}(u) = \int_{-\infty}^{+\infty} f(x)g(x-u)\,\mathrm{d}x \ . \qquad (4.42)$$

A simple 1-dimensional example of autocorrelation is shown in Fig. 4.10(a) and (c). ((b) and (d) refer to §4.7.1.) The reader should draw a few

Fig. 4.10 — Example of:

autocorrelation – (a) (c);
Wiener–Khintchine theorem – (c) (d).
T denotes Fourier transformation.

sketches to show that when a second rectangle function, identical to the one in (a), is shifted by various amounts along the x-axis, the area under the product for each shift builds up the result shown in (c).

The definition of correlation is often expressed in a slightly different form from the above, as follows:

$$c(x) = f(x)\odot g(x) = \int_{-\infty}^{+\infty} f(x')g(x'-x)\,\mathrm{d}x' \qquad (4.43)$$

and similarly for autocorrelation. In this formulation x is the shift and x' is the dummy variable for integration.

Correlation is therefore like convolution (cf. Eqn (4.24)) but differs in that g is shifted but not reversed (folded). The symbol $\odot$ is used fairly widely to distinguish correlation from convolution, though unfortunately there are no universally adopted symbols for these operations.

When the functions are complex it is customary to define the **complex cross-correlation function** as

$$\gamma_{fg}(x) = f(x)\odot g^*(x) = \int_{x'=-\infty}^{x'=+\infty} f(x')g^*(x'-x)\,dx' \qquad (4.44)$$

and the **complex autocorrelation function** as

$$\gamma_{ff}(x) = f(x)\odot f^*(x) = \int_{x'=-\infty}^{x'=+\infty} f(x')f^*(x'-x)\,dx' \ . \qquad (4.45)$$

(Note that if the asterisk is associated with the other function the conjugate result is obtained.)

For $f(x)$ and $g(x)$ complex, the difference between convolution and correlation referred to above can be summarized as follows:

$$f(x)\odot g^*(x) = f(x)\circledast g^*(-x) \ . \qquad (4.46)$$

It follows that, unlike convolution, correlation is not commutative i.e.

$$f(x)\odot g^*(x) \neq g^*(x)\odot f(x) \ . \qquad (4.47)$$

For some purposes it is convenient to normalize correlation functions and this is done by dividing the expressions in Eqns (4.44) and (4.45) by the central value of the correlation, i.e. by the value for $x = 0$ (zero shift). This gives

$$\text{Normalized } \gamma_{fg}(x) = \frac{\displaystyle\int f(x')g^*(x'-x)\,dx'}{\displaystyle\int f(x')g^*(x')\,dx'} \qquad (4.48)$$

$$\text{Normalized } \gamma_{ff}(x) = \frac{\displaystyle\int f(x')f^*(x'-x)\,dx'}{\displaystyle\int f(x')f^*(x')\,dx'} \qquad (4.49)$$

$$= \frac{\displaystyle\int f(x')f^*(x'-x)\,dx'}{\displaystyle\int |f(x)|^2\,dx'} \qquad (4.50)$$

where the integrations are from $-\infty$ to $+\infty$.

4.7.1 Autocorrelation theorem (Wiener–Khintchine theorem)

This theorem states that the Fourier transform of the autocorrelation of a function $f(x)$ is the squared modulus of the transform of $f(x)$.

Taking the general case, if the transform of a complex function is $F(u)$ the theorem means that the transform of $|F(u)|^2$ is the complex autocorrelation of $f(x)$.

Using T to denote Fourier transformation, we have

$$T[|F(u)|^2] = \int_{-\infty}^{+\infty} |F(u)|^2 \, e^{2\pi iux} \, du$$

$$= \int_{-\infty}^{+\infty} (F(u)F^*(u)) \, e^{2\pi iux} \, du$$

$$= f(x) \circledast f^*(-x)\dagger \tag{4.51}$$

whence, using Eqn (4.46),

$$T[|F(u)|^2] = f(x) \odot f^*(x) \tag{4.52}$$

i.e.

$$F(u)F^*(u) \xrightarrow{T} f(x) \odot f^*(x) \tag{4.53}$$

This is illustrated in Fig. 4.10 for the real function $f(x)$.

Changing to time and frequency gives

$$T[|F(v)|^2] = f(t) \odot f^*(t) \tag{4.54}$$

where $|F(v)|^2$ is described as the **power spectrum** as a function of (*temporal*) *frequency*. In Eqn. (4.52) $|F(u)|^2$ is likewise referred to as a **power spectrum** in relation to *spatial frequency*. Applications of both await us in the chapters ahead.

---

† The negative sign in $f^*(-x)$ arises from Fourier transformation of a conjugate:

$$\int_{-\infty}^{+\infty} F^*(u) \, e^{2\pi iux} \, du = \left[\int_{-\infty}^{+\infty} F(u) \, e^{-2\pi iux} \, du \right]^*$$

i.e. $T[F^*(u)] = f^*(-x)$.

To appreciate more fully why these are called power spectra we start on familiar ground by recognizing $F(v)$ as the amplitude distribution of the frequency spectrum of a temporal function $f(t)$, where $F(v)$ and $f(t)$ are a Fourier transform pair: the same applies to $F(u)$ and $f(x)$ in the space domain. Now if $f(t)$ is the amplitude of e.g. a wavemotion, as a function of time, then the total energy of the wavemotion with respect to time can be expressed as proportional to

$$\int_{-\infty}^{+\infty} |f(t)|^2 \, \mathrm{d}t \ ,$$

since energy is proportional to (amplitude)^2. To show that this is the same as the equivalent expression in terms of frequency, with $|F(v)|$ described as the '**power spectrum**' as above, we proceed by writing

$$|f(t)|^2 = f(t)f^*(t) \tag{4.55}$$

$$= f(t)[T\{F(v)\}]^* \ , \tag{4.56}$$

where T denotes Fourier transformation as usual. We then have

$$\int_{-\infty}^{+\infty} |f(t)|^2 \, \mathrm{d}t = \int_{-\infty}^{+\infty} f(t)\left[\int_{-\infty}^{+\infty} F^*(v) \, \mathrm{e}^{2\pi i v t} \, \mathrm{d}v\right] \mathrm{d}t \ . \tag{4.57}$$

Changing the order of integration,

$$\int_{-\infty}^{+\infty} |f(t)|^2 \, \mathrm{d}t = \int_{-\infty}^{+\infty} F^*(v)\left[\int_{-\infty}^{+\infty} f(t) \, \mathrm{e}^{2\pi i v t} \, \mathrm{d}t\right] \mathrm{d}v \ , \tag{4.58}$$

i.e.

$$\int_{-\infty}^{+\infty} |f(t)|^2 \, \mathrm{d}t = \int_{-\infty}^{+\infty} |F(v)|^2 \, \mathrm{d}v \ , \tag{4.59}$$

which is the relationship we sought. In the space domain

$$\int_{-\infty}^{+\infty} |f(x)|^2 \, \mathrm{d}x = \int_{-\infty}^{+\infty} |F(u)|^2 \, \mathrm{d}u \ . \tag{4.60}$$

In words, the relationship tells us that the integral of the squared modulus of a function is equal to the integral of the squared modulus of its spectrum.

These are examples of **Parseval's theorem** which, in its generalized form is expressed as

$$\int_{-\infty}^{+\infty} f(x)g^*(x) \, \mathrm{d}x = \int_{-\infty}^{+\infty} F(u)G^*(u) \, \mathrm{d}u \ , \tag{4.61}$$

which is the same as the equation we have derived, when $f = g$.

In the form of Eqn (4.60), Parseval's theorem has been described as an expression of the principle of the conservation of energy.

PROBLEMS

4.1 Use a phasor diagram to show how there is a phase shift in the double-slit diffraction pattern when the slits in Fig. 2.08 are moved through a distance $D/2$ in the x-direction. Compare the result with the analytical treatment in §4.3.

4.2   Sketch 'by inspection' (a) the autocorrelation function of a row of five equally spaced dots, (b) the autocorrelation of the letter A stylized by the dot pattern

$$
\begin{array}{c}
\cdot \\
\cdot \; \cdot \\
\cdot \; \cdot \; \cdot \\
\cdot \qquad \cdot
\end{array}
$$

and (c) the cross-correlation of two rows of dots as in (a), positioned at right angles to each other. Comment as fully as you can on each result.

4.3   The Fourier transform of an array of δ-functions

$$ f(x) = \sum_N \delta(x - x_N) $$

is given by

$$ F(u) = \sum_N e^{2\pi i u x_N} . $$

Use this to obtain an expression for the 'grating term' in the diffraction pattern given by three apertures located at $x = -D,\ 0,\ +D$. Sketch the intensity distribution in the pattern for this term and indicate how it is modified by the 'single-aperture' term if the latter is for a rectangle function of width a.

4.4   In Fig. 4.05(e) the shaded area gives the total value of the output at x_1, i.e. $h(x_1)$, when an input $f(x)$ is convoluted with a smearing function $g(x)$. Show that the ordinate at say x_2 in this shaded area gives the contribution to the output at x_1 made by the smearing of the input at x_2.

4.5   A signal $s(t)$ defined by
$$
\begin{aligned}
s(t) &= 0 & & t \leqslant 0 \\
&= t/\Delta & & 0 < t \leqslant \Delta \\
&= 0 & & t > \Delta
\end{aligned}
$$

is applied to a system whose response is

$$h(t) = \frac{1}{\Delta} e^{-t/\Delta},$$

which is typical of a number of simple physical systems.

(a) Use the convolution integral to determine the system output.
(b) Perform the convolution graphically by dividing the signal into strips, each of which sets up the impulse response, using the value $\Delta = 1$ second.

5

Optical imaging and processing

5.1 INTRODUCTION

The major aspects of optical image formation and processing under coherent and incoherent illumination are outlined in this chapter. They involve Fourier transforms, alone or in association with convolution and correlation. Attention must be drawn without delay, however, to the fact that their importance is not restricted to the handling of data that are optical in origin. There are numerous instances today where optical processing is used even when the initial data are not optical. The basic reason for this is the way in which the same mathematical operations that apply to most optical systems also often apply to electrical communications systems. The optical analogue is attractive because it has the advantage of 2-dimensional display and parallel processing; and it is being made increasingly practicable by the development of electrical-to-optical interface devices, in conjunction with the computer. When, for some reason, practical optical techniques are not used, the computer can be used alone to enable the same fundamental principles to be employed for 'digital imaging' and processing.

Our purpose in this chapter is, however, solely to deal with the essentials of how Fourier transforms, convolution and correlation, already introduced in the optical context in previous chapters, form much of the basic tool-kit for optical imaging and processing. Some notes on applications in the field of communications are to be found in Appendix I.

Before embarking on these topics in detail it will be helpful to have a descriptive preview of the sections (5.2 and 5.3) on optical image formation itself. **Optical processing**, as opposed to imaging, is concerned with intervention in the process in various ways and for various purposes, and that is the subject of §5.5.

There are two models of optical imaging that are useful today, each with applications to which it is particularly well suited. Both were touched on in

Chapter 1 (see especially Fig. 1.08) and both have been mentioned to varying extents in the intervening chapters.

The *first model* places emphasis on the overall diffraction (scattering) of light from an object when the conditions are at least partially coherent, and the way in which that light is recombined to form an image. The Fourier aspects of the first part of this are already familiar from Chapters 3 and 4. In §5.3 we consider this again, particularly in relation to the second stage of image formation. This model was originally formulated essentially qualitatively by E. Abbe in 1873 to deal with his experimental observations on the imaging of periodic objects under the microscope. In today's terminology he found that with the illumination methods normally used in optical microscopy, image formation is not the completely incoherent process sometimes assumed; in fact it can be nearly coherent with some modern systems.

For a coherently illuminated object the quantitative identification of diffraction orders with the terms of the Fourier series analysis describing its structure, and the way an imaging lens recombines them as a Fourier synthesis to form an image, were reported in some detail by A. B. Porter in 1906. (It does not seem to be widely appreciated that a significant step in that direction had already been taken by Rayleigh in 1874, and certainly in his paper of 1896.)

The extension of the Abbe–Porter interpretation to include the analogous role of Fourier transforms in the coherent imaging of non-periodic objects was made mainly by X-ray crystallographers by the early 1940s.

The value of the approach offered by this model is its sensitivity to the way in which spatial frequencies in an object 'structure' (periodic or non-periodic) are conveyed by diffracted wavefronts and reconstituted to form an image. Using coherent illumination it leads to methods for intervention in the diffraction plane (the spatial frequency plane) so that the formation of the image can be controlled by 'filtering'. This is one aspect of optical processing; others are mentioned in §5.5.

The *second model* of image formation, which we consider in §5.2, is applicable to both coherent and incoherent illumination conditions. Here again Rayleigh made an important contribution (1896), this time with influences from the earlier work of Airy and Helmholtz. The model visualizes an image as the combination of the Airy patterns (or more complicated patterns if aberrations are present) that an optical system would separately produce with the light leaving every individual point in the object. As we noted in §1.3, the Fraunhofer diffraction patterns given by aperture masks are formed in the plane where the image of the light source is formed, i.e. the plane 'conjugate' to the light source. In the present instance, in the model of imaging developed by Rayleigh, each point in an object being imaged is itself considered as a point source of light. The wavefronts from it are limited in their entry into the imaging system by the finite aperture of the imaging lens, and the diffraction pattern of that aperture is formed in the plane conjugate to the plane containing the point source, i.e. in the image plane of the object points as light sources (Fig. 1.08). Each point of the object is therefore imaged not as a point but as the Airy pattern of the aperture of the imaging lens (or more complicated pattern if aberrations are present).

If the illumination is incoherent the Airy intensity patterns due to all the object points are simply additive. If it is coherent there is interference and mathematically one deals with the combination of the complex-amplitude Airy patterns. Rayleigh dealt with both extremes. By considering an object in the form of a row of equally spaced point sources he used a Fourier series to represent the summation of the regularly spaced Airy intensity patterns forming the image according to this model. For coherent conditions he showed that the terms of the series correspond to the diffraction orders of the Abbe model referred to above.

An important development of this model of image formation is due to Duffieux who, in his book *L'Intégrale de Fourier et ses applications a l'optique*, acknowledged his indebtedness to 'Michelson et surtout Lord Rayleigh' as the founders of the use of Fourier methods in physical optics. We shall read a great deal more about Michelson in Chapter 7. And of Rayleigh, that most versatile scientist who made major contributions in nearly every branch of physics, he gained the Nobel Prize in 1904 for physics, for his investigations on the densities of the more important gases and for his discovery of argon!

Duffieux appears to have been the first formally to express the image of a continuous object distribution as a convolution (he used the German *'faltung'*) of that distribution with what we call the **point-spread function**, or **impulse response** of the system (the Airy pattern in the simplest case), using complex amplitudes or intensities as appropriate for the illumination conditions. Application of the convolution theorem then showed him that the spatial frequency spectrum of the image is the product of the frequency spectrum of the object distribution and that of the response of the system. The optical system could therefore be considered as 'transferring', to different extents, the spatial frequency components of the object to the image plane. This work gave us one of the most important concepts in this subject today, namely that of an optical imaging system as having a **transfer function** (Duffieux's *'facteur de transmission'*) for each frequency entering it. When it is applicable — and we shall come to that in a moment — it provides an invaluable approach to matters concerned with the design and analysis of optical systems. Its virtues may be seen as complementary to those of the Abbe–Porter model.

Another influence in establishing this **convolution model** of imaging has come from the work of Schade (1948), Elias *et al.* (1952) (see also Elias, 1953) and others who have pointed to the link that exists between these aspects of optics and the ideas and methods that had been developed in connection with the analysis of electrical networks and other communications problems of a **linear systems** type. A short digression here into this other area is justified by the valuable cross-fertilization of ideas that has resulted from the recognition of the way in which the two share the same fundamental principles.

In the general sense a '**system**' may be defined as any device that 'maps a set of input functions on to a set of output functions'. And as noted in §4.4.1, in slightly different terms, a system is '**linear**' if its response to a number of input signals is the sum of the responses obtained when the signals are applied separately. In the optical context a system is linear if, as is commonly the case, spatial frequencies are transmitted through the system with alteration only of their amplitude and phase, not their frequencies. (**Non-linear systems** are outside the scope of this book.)

In the design and analysis of linear electrical networks one method had been to consider an output in just the same way as we have described above for optical image formation, viz. a convolution of the input (expressed as a series of impulses of changing magnitude) with the unit impulse response of the system. The integrations involved when the effects of different filters were being investigated, became very complicated. Even more difficult was the deconvolution needed in the design of filters required to produce specified outputs from predetermined inputs. It was the application of the convolution theorem that produced the much needed simplification in many instances. From the theorem it follows that the temporal frequency spectrum of the output from a linear electrical network system is simply the product of the frequency spectrum of the input and the frequency spectrum of the unit impulse response of the system (its transfer function). Integration in the time domain is replaced by the simpler operation of multiplication in the frequency domain. Furthermore, the overall frequency response of several filters in tandem is simply the product of their individual transfer functions. Small wonder that it was said that if network theory had been limited to the time domain approach, it would never have advanced.

However, the straightforward use of the convolution theorem as described above, whether in communication systems (Appendix I) or in optical imaging systems, necessitates an additional requirement, namely that of **invariance** (or 'stationarity'). Strictly speaking, this means that in an electrical network, for example, the response to a unit-impulse must be independent of the time at which that input is applied — i.e. it is a **time-invariant** system. Likewise, in an optical imaging system the image of a point object — the point-spread function — must be the same over the entire field; it must be a **space-invariant** system (cf. §4.4.1). Early in the following section we will dicuss the implications of this in optical image processing.

Electrical engineering has had other influences on optics, notably arising from the work initiated and developed by Wiener (1949) and Shannon (1949) on **statistical network theory** and **information theory**. This has led, for example, to the use of correlation functions in the real-space processing of optical images, some aspects of which are introduced in §5.5.

5.2 INCOHERENT OPTICAL IMAGING

The description in the previous section of image formation as a convolution can usually be applied to the normal use of a telescope or camera etc., as a convolution of intensities. For most photographic, television, and other systems it has been found that there is good agreement between experimental measurement and calculation based on the assumption that illumination under normal conditions is essentially incoherent (Barnes, 1971). Happily, there are a number of advantages in this, including the opportunity of using TV pictures, LED displays, etc., as inputs to processing systems.

When optical aberrations are present in the system, however, the unit impulse response of the system, which is here the point-spread function of the system (§2.3), may well be different for different points in the object field. Such variations could, as we have seen, invalidate the use of the convolution theorem. Fortunately, when a system is well corrected the residual effects of aberrations are reasonably constant over the area within which the image of any point in the object field is of significant intensity. The system is then said to be **isoplanatic** (analogous to it being 'achromatic' when the positions of image points are independent of the wavelength of the light in the wavelength range concerned). Under those circumstances the system is for practical purposes space-invariant (see previous section), and the convolution theorem can be used in a straightforward way.

As indicated in Fig. 5.01, though purely symbolically in 1-dimension, the

Fig. 5.01 — Spread and transfer functions. Schematic diagram of functional relationships for incoherent illumination. T, Fourier transform; PSF, point-spread function: OTF, optical transfer function; ⊛ convolution; × multiplication.

application of the convolution theorem gives the frequency spectrum of the image intensity distribution as the product of the frequency spectrum of the object intensity distribution and the Fourier transform of the PSF. The transform of the PSF is the **optical transfer function** (OTF) of the system.

This means that every sinusoidal component in the object distribution is 'transferred' to the image without alteration of its frequency: only its intensity and phase are subject to alteration by the system. Consider the intensities first. Figure 5.02(a) represents an object intensity distribution containing a single sinusoidal component of some specified spatial frequency. The 'strength' of this component is expressed in terms of the modulation M, given by

$$M = \beta/\alpha \tag{5.01}$$

(cf. definitions in §1.1 and Eqn (1.05)).

The ratio of the modulation in the image to that in the object is the modulation transfer *factor* for that frequency: effectively an index of the efficiency of the imaging system at that frequency, it is usually less than unity (except for zero frequency). Expressed as a function of frequency over the appropriate range it comprises the **modulation transfer function** (MTF) of the system.

Any frequency-dependent phase changes introduced by the system would be manifest in the image as lateral shifts of the sinusoidal components comprising the image, as shown in Fig. 5.02(b). Complementary to the MTF, there is therefore a **phase transfer function** (PTF) though this is often negligible.

The two functions comprise the optical transfer function (OTF) (or **frequency response**) of the system as follows

$$(\text{OTF}) = (\text{MTF}) \, e^{i(\text{PTF})} . \tag{5.02}$$

5.2.1 Determination of transfer functions

(i) *Calculation*

Several methods are available for calculating the OTF of an optical system from its design data. For the contribution due to aberrations one method is to compute the passage of a large number of rays through the system from a single object point. By spacing the rays uniformly across the aperture of the lens, the density distribution of resulting points in the image plane gives the intensity distribution that comprises the point-spread function. Fourier transformation then gives the **geometrical** OTF of the system. If a system is free from aberrations the geometrical OTF would be unity for all frequencies; every object point would be imaged as a point. Correction for diffraction is made by multiplying this geometrical transfer function by the transfer function for the equivalent diffraction limited system, i.e. the system free from all imperfections.

Other methods involve the detailed computation of the aperture function, including the effects of aberrations. This is a complex amplitude distribution over the aperture and we will denote it by $f(x)$ as for aperture functions in previous chapters. Its Fourier transform, $F(u)$, is the complex amplitude diffraction pattern of the image of a point source. The squared modulus of this is the PSF, and the Fourier transform of that is the OTF. This is illustrated symbolically in 1-dimension in Fig. 4.10 for the familiar

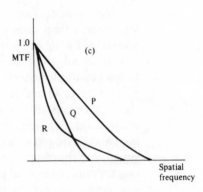

Fig. 5.02 — Transfer functions (incoherent illumination).
   (a) Modulation of an individual frequency component is given by β/α;
   (b) associated phase shift;
   (c) examples of modulation transfer functions.

example of $f(x)$ as a simple rectangle function. The calculation route just described is (a) → (b) → (d) → (c) in that figure.

Alternatively, writing

$$OTF = T\{|F(u)|^2\}$$

where T denotes Fourier transformation, and using the autocorrelation theorem (§4.7.1), gives

$$OTF = f(x) \odot f^*(x) \qquad\qquad (5.03)$$

i.e. the OTF is given directly by calculating the autocorrelation of the aperture function. In Fig. 4.10 this is the direct route (a) → (c).

(ii) *Experimental*

Various methods, some quite complicated, are employed in the experimental determination of transfer functions. Only a simple method, chosen because it re-emphasizes the meaning of what is involved, need be mentioned here. In this the MTF is determined for one frequency at a time. A series of object screens is used, each bearing a sinewave pattern of different spatial frequency; they look rather like an out-of-focus shadow of a row of parallel, evenly spaced knitting needles. The ratio of the modulation in the image to that in the object is determined for pattens of spatial frequencies covering the necessary frequency range. Frequency-dependent phase shifts, comprising the PTF, are given by the relative lateral positions of the image and object bands at each calibration frequency. By its nature this method gives the 'line-spread function' (LSF) rather than the point-spread function (PSF).

These transfer functions provide a much more informative appraisal of a lens system than does just a measurement of its resolution limit. This is illustrated by the MTF curves in Fig. 5.02(c). Curve P corresponds to a lens free from all aberrations, the contrast ratio decreasing with increasing frequency until it reaches zero at the resolution limit of the lens (cf. Fig. 5.01). Curves Q and R are representative of lenses with aberrations; they show that whilst R has a frequency limit superior to Q, it gives an image contrast (modulation) that is inferior at lower frequencies; choice between the two can be made according to the type of application. Though greatly superior to the older and rather misleading measurement of resolution limit as a criterion of optical performance, optical transfer functions are not the complete answer to the problem of assessing performance, especially if the eye is involved in the final imaging. The eye is a relatively poor imaging system but it is associated with complicated data-processing in the retina and brain. This makes the prediction and determination of its overall response in any particular situation very difficult.

5.3 COHERENT OPTICAL IMAGING

Ernst Abbe's experimental work in the 1870s on improving the performance of microscope objectives laid the foundation of the approach used today in considering coherent imaging. Working at the University of Jena in collaboration (later in partnership) with Carl Zeiss at the latter's optical works, Abbe found that microscope objectives made with the most careful correction of aberrations gave resolution inferior to less carefully corrected objectives of larger aperture. In experiments with periodic specimens such as insect scales and diatom skeletons he showed that the influence of a large aperture is associated with diffraction at the specimen itself (− evidence of some coherence in the illumination conditions).

He demonstrated how the diffraction maxima formed in the back focal plane of an objective contribute to the formation of the image, the higher orders (or higher spatial frequencies as we now regard them) controlling the fine detail in the image. Abbe had thus introduced wave theory ('wave optics') into instrumental optics, previously the exclusive province of geometrical (or 'ray') optics.

The interpretation of Abbe's work in terms of Fourier series was described in an important paper by A. B. Porter in 1906. And in an elegant series of experimental demonstrations he also showed their physical reality. For example, he showed the effects on the image of a periodic object when different combinations of diffraction orders leaving the object are prevented from contributing to the image.

Subsequently extended by the application of Fourier transform methods, to deal with the imaging of non-periodic objects, the Abbe approach has been put to use in a number of outstandingly important ways. As already mentioned, these mainly depend on the spatial frequency aspects of Fraunhofer diffraction and the accessibility of the diffraction pattern, mathematically and experimentally, if coherent conditions are used.

5.3.1 Periodic objects

Consider the 1-dimensional multiple-slit transmission grating of earlier chapters, acting as an object being imaged by a lens (Fig. 5.03). Wavefronts

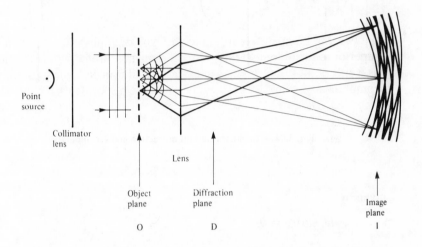

Fig. 5.03 — Abbe's treatment of image formation in the microscope (after Bragg, 1929).

constituting the various diffraction maxima are brought to separate foci in the back focal plane of the lens, and it is the light that passes through these foci in the diffraction plane that combines in the image plane to give an optical reconstruction of the object: for the moment we will assume the lens to be perfect and of infinite aperture. Note that, as indicated by the two heavy lines from the object, there is no conflict between the wave description and the ray-tracing method of geometrical optics.

Alone, *any* pair of the foci produces a set of sinusoidal bands in the image plane. This is reminiscent of Young's experiment (§1.1) where the pair of apertures act in the same way. In this sense, image formation can be thought of as a double process of diffraction — an approach put forward by F. Zernike in about 1935.

To see how an image is built up in this way, with the type of object grating envisaged in Fig. 5.03, we will consider the contributions to the image made by the pairs of diffracted beams comprising the various orders of diffraction.

Figure 5.04 shows, schematically, the formation of the nth order pair of diffracted beams leaving the object, and how in the image plane they interfere to produce a harmonic distribution of illumination with repeat D'_n

Fig. 5.04 — Image formation: diffraction and recombination.

given by

$$D'_n \sin \theta'_n = \lambda \ . \tag{5.04}$$

The condition for the formation of the nth order diffraction maxima at the object is

$$D \sin \theta_n = n\lambda \ , \tag{5.05}$$

whence
$$D'_n = \frac{D}{n} \left(\frac{\sin \theta_n}{\sin \theta'_n} \right) \ .$$

The term in brackets is a constant related to the magnifying power of the lens. So we have

$$D'_n \propto D/n \ . \tag{5.06}$$

The first-order pair of maxima from the object interfere in the image plane to give a simple-harmonic variation of illumination that corresponds just to the basic repeat of the grating: this repeat is the bare minimum of information about the object, with no fine detail of its optical structure. Each pair of successively higher-order maxima adds a harmonic of shorter repeat ($\propto D/n$) to the total illumination that forms the image: the full detail of the image is built up by what is clearly recognizable as Fourier synthesis. Now in §3.4.1 it was shown that the diffraction maxima themselves comprise the Fourier analysis of the object grating, and reference was made to the diffraction plane being described as the Fourier plane. Therefore the **image**

formation process in the example we are considering may be regarded as a **double Fourier process**, with the diffraction pattern as a Fourier analysis of the grating, and the image as a Fourier synthesis of the Fourier analysis. This is particularly evident when one invokes the **principle of reversibility**. All the diffraction orders that build up the image as a summation of harmonics return to the object grating where they recombine in the same way to form the original illumination distribution (the aperture function) at the grating.

It is important to appreciate that this double Fourier description is in complete accord with Zernike's double-diffraction interpretation; they are different ways of expressing the same thing.

For perfect imaging, an infinite Fourier synthesis would be required, necessitating not only the generation of an infinite number of diffraction orders but also their entire admission into the optical system. Both are clearly impossible. Eqn (5.05) shows how the values of D and λ impose a limit to the number of possible diffraction maxima that can be produced; and any objective of course has a finite aperture.

At the other extreme, if conditions allowed only the zero-order and first-order pair of diffraction maxima to enter the objective only the basic repeat, D, in the object grating would be resolved in the image; and then, as noted above, only as a simple sinusoidal variation of illumination. Eqn (5.05) shows that on this basis the smallest object repeat that is resolvable is given by

$$D \sin \theta = 1\lambda$$

i.e. $$D = \lambda/\sin \theta .$$ (5.07)

This gives the **resolution limit** as approximately the wavelength of the illumination used. (The resolution limit can be reduced below that described above since the combination of the zero-order with only one of the first-order maxima is sufficient to form an image of the basic grating repeat. By using convergent illumination it can be arranged that these just enter opposite sides of the aperture of the objective, and the resolution limit is then $\sim \lambda/2$. This minimal need for the zero-order and one first-order maximum is the **Abbe principle**.)

Exactly the same general findings apply in the case of a 2-dimensional grating, i.e. a grating in the form of a 2-dimensional lattice at each point of which is an identical aperture (§2.6). In the case of a fabricated grating each 'aperture' can, for example, be a pinhole or a group of pinholes. The diffraction pattern is then in the form of a 2-dimensional lattice of spots of illumination, the diffraction order of each spot being specified by two integers (cf. the three integers in Eqns (2.18) for a 3-dimensional grating). Recombination, the second stage in the formation of the image, is completed in exactly the same way as for the 1-dimensional example.

In §5.3.3 the experimental observation of some of these effects is described.

5.3.2 Non-periodic objects

The treatment in the previous section can be extended to deal with non-periodic objects because *discrete* orders of diffraction are not a prerequisite. A non-periodic object can be regarded as equivalent to one aperture of grating, and we know that under those circumstances a Fourier transform instead of a Fourier series is involved. The diffraction pattern in the focal plane of the lens is one of continuous scattering, with a directional variation of amplitude and phase depending on the aperture function: it is the Fourier transform of the amplitude distribution function of the object (cf. the lens as a 'Fourier transformer' in §4.2). Recombination of this pattern in the image plane is a summation of fringes produced by pairs of diffracted beams (at $\pm\theta'$ in Fig. 5.04) but with a continuous range of fringe spacings and orientations. Image formation can then be described as a double Fourier transformation process. This description is commonly applied to both periodic and non-periodic objects since even the former are of finite extent and, as noted in §4.2, it is very convenient to be able to speak of an image as being the transform of a diffraction pattern regardless of the exact nature of the object. We have already used the idea in §4.5.

5.3.3 Illustrations: optical transforms

An optical diffractometer set-up for demonstrating and using the concepts described above is shown in a simplified form in Fig. 5.05. A helium–neon

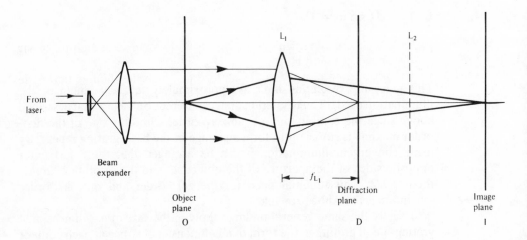

Fig. 5.05 — Basic arrangement for observing optical transforms.

laser is commonly employed, with a beam expander to provide illumination with near-perfect coherence (temporal and spatial) across the whole plane-wavefront at O where object masks are placed. The diffraction pattern (Fourier transform) of a mask at O is formed at D, the focal plane of the objective L_1; and the image (the double transform) of O is formed at I. In practice, to obtain diffraction patterns of reasonable size, L_1 is a long focal length lens or equivalent (e.g. a telephoto combination). A second lens, L_2, (only its position is shown (dashed line) in the figure) is then required for the formation of a real image at a reasonable distance from the object mask.

Photographic records of diffraction patterns and images are, by their nature, records of intensities (except in holography). In the following examples the illustrations are therefore described as **optical transforms**, rather than Fourier transforms which are statements of amplitude and phase. They are mainly taken from the definitive collection of books on optical transforms by H. Lipson and his colleagues (see the Bibliography).

(i) *Diffraction by object masks*

The sequence in Fig. 5.06 illustrates some of the main points that have been mentioned concerning the role of diffraction in the first stage of image formation in coherent light. Each example shows an object mask and the optical transform it gives.

(a) *Object mask*: a 2-dimensional 'point' lattice of pinholes. It can be regarded as a lattice array of δ-functions. The mask contained very many more repeats than are shown here.
Transform: a reciprocal lattice — its geometry and dimensions are reciprocally related to those of the object lattice. Intensities are roughly constant, as expected for small pinholes approximating to δ-functions (i.e. the Airy disc is large).

(b) *Object mask*: a single aperture-unit consisting of a non-repeating, irregular group of holes.
Transform: a non-repeating pattern in which the illumination varies with direction in a way determined by the arrangement and sizes of the holes comprising the object mask.

(c) *Object mask*: the convolution of (a) and (b). As in (a), the mask contained very many more repeats than are shown here.
Transform: the product of the individual transforms of (a) and (b). This illustrates the convolution theorem. It also illustrates how the pattern given by a grating is a sampling of the pattern given by the unit comprising the grating (i.e. its transform). The sampling is in directions determined by the dimensions of the grating lattice (and, of course, by the wavelength of the light).

(d) *Object mask*: a double-aperture 'grating' which is a convolution of the single aperture unit in (b) with two δ-functions.
Transform: again the product of the two separate transforms. In this instance the transform of one unit is multiplied by that of a double-aperture grating. Note that the $\cos^2$ fringes of the latter are perpendicular to the line joining the two 'apertures' (as in Young's experiment): their spacing is reciprocally related to the separation of the apertures.

Fig. 5.06 — Optical transforms.
Each transform is accompanied beneath by its mask, except in (a) and
(c) for which the masks contained many more repeats than are shown
here.
(Adapted with permission, from Taylor and Lipson (1964), *Optical transforms*. Bell,
London.)

(e) *Object mask*: a convolution of the single aperture unit in (b) with four δ-
functions that define the grating lattice of (a).
Transform: the transform of the aperture unit is now multiplied by two
sets of $\cos^2$ fringes that define the reciprocal lattice of (a). The sampling
aspect is again demonstrated.

Elegant examples of the use of optical transforms are to be found in X-
ray crystallography where, as noted in Chapter 2, the formation of images of

atoms cannot be completed directly because there are no lenses that can be used to recombine the diffracted X-rays. Stopped short at the diffraction stage, only intensities are recorded and Fourier summations cannot be performed numerically or experimentally because of the lack of phase data. In the formative years of the subject W. L. Bragg played a key role in pioneering and developing the use of Fourier and optical principles in understanding and dealing with this and other problems in X-ray crystallo-graphy. Whilst the development of the digital computer has led to computa-tional methods for overcoming the 'phase problem' Bragg's work made an important contribution to the wider field of optical processing. For the main literature on the development and applications of optical methods to X-ray diffraction the reader should consult the books referred to at the beginning of this section.

An example of the analogy between X-ray diffraction by a crystal and the first stage of optical image formation of a grating object is shown in Fig. 5.07.

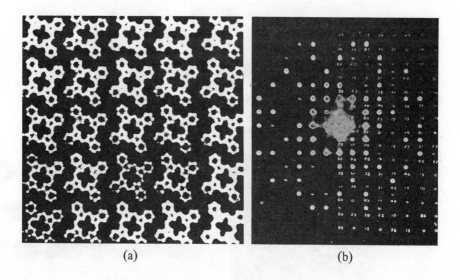

(a)   (b)

Fig. 5.07 — The first stage of image formation.
(a) Mask representing a 2-dimensional projection of the crystal structure of phthalocyanine;
(b) optical diffraction pattern given by (a).
(Reproduced by permission from *Nature*, **154**, 69. Copyright © 1944 Macmillan Journals Limited.)

(a) is part of an optical mask representing a 2-dimensional view, in a particular direction, of the crystal structure of phthalocyanine; (b) is the optical diffraction pattern it gives (Bragg, 1944). Not only does (b) agree with the geometry of the X-ray data but the intensities also agree with the experimentally observed X-ray intensities — indicated by the numbers alongside the spots in the figure. Historically, this was a method for determining an unknown crystal structure by making 'trial' masks based on chemical and other considerations. The method was made simpler in later developments of the technique (referred to at the end of (ii)) when it was shown that the basic unit and only three repeats are sufficient as a mask since these define the lattice on which the 2-dimensional projection of a crystal is based (Hanson and Lipson, 1952). This is illustrated in Fig. 5.06. In (c) a large number of repeats (many more than shown in the figure) of the basic unit were used as the mask, whereas in (e) only four units defining the lattice repeats were used. Comparison of the optical transforms shows that (e) is quite adequate, compared with (c), in giving a sampling of the transform of the single unit (b) at the reciprocal lattice points (the transform) of the basic lattice (a).

(ii) *Recombination*

Figure 5.08 is another example from X-ray crystallography, this time of the

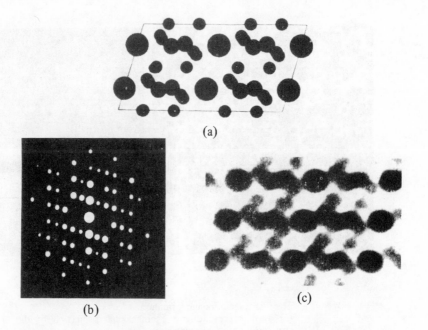

Fig. 5.08 — The second stage of image formation.
(a) Representation of a 2-dimensional projection of the crystal structure of diopside, $CaMg(SiO_3)_2$; (b) optical mask representing the
corresponding X-ray diffraction pattern; (c) optically reconstructed
image using (b).
(Reproduced by permission from *Nature*, **143**, 678; **149**, 470. Copyright © 1939 and
1942 Macmillan Journals Limited.)

second stage of image formation, the recombination stage. (a) is a diagram-matic 'view' down one axis of the crystal structure of the mineral diopside, $CaMg(SiO_3)_2$, which W. L. Bragg used on several occasions to illustrate the optical principles underlying the 'X-ray analysis' of crystal structure (Bragg, 1929, 1939, 1942, 1944). The atoms are represented by black discs, the larger ones representing Ca and Mg atoms which are superimposed in this view, the smaller ones representing the Si and O atoms (which form chains running through the structure). In (b) we have a mask containing small holes to represent the experimentally observed X-ray diffraction data; the areas of the holes are made proportional to the square-roots of the X-ray intensities. (c) is the image formed as the optical transform of (b) and it is a very realistic match to (a). The recombination has succeeded because nearly all the X-ray diffracted beams have the same phase, being largely determined by the strongly scattering Ca and Mg atoms which are at centres of symmetry. In examples where different phases are involved this can be allowed for by placing a piece of mica in front of each hole and rotating the pieces individually to produce the appropriate phase shifts. (Contrary to first impression this method does not require the use of polarized light in the optical reconstruction of the image.)

That such images can be regarded as composed of sinusoidal harmonics produced by pairs of diffraction maxima, as described in §5.3.1, had earlier been demonstrated by Bragg. Figure 5.09 shows sets of sinusoidal bands like

Fig. 5.09 — Sinusoidal bands of light and shade formed in the image plane by pairs of diffraction maxima. (Reproduced by permission, from Bragg (1933), *The crystalline state*. Bell, London.)

those produced separately by pairs of different diffraction maxima, their phases manifest by the positions of their 'crests' and 'troughs' with respect to the origins at the corners of the unit cell. Bragg exposed a sheet of photographic paper successively to artificially made patterns like these to represent the diffraction effects produced by the mask in Fig. 5.08(b). Keeping exposures down to a level such that the general level of photographic blackening did not build up too much he obtained a picture like that in Fig. 5.08(c).

• • • • •

The use of X-rays in the first stage of image formation followed by the use of visible light and a lens to complete the second stage became known (Buerger, 1950) as **2-wavelength microscopy**, although Bragg preferred 'X-ray microscopy'. It was to be a major source of inspiration to D. Gabor in his investigation of holography (§5.4) and thence to the use of visible light in processing electron microscope images, developed by A. Klug and his colleagues (§5.5.1).

As mentioned above, the lack of experimentally determined X-ray phase data, required for image formation to be completed by optical reconstruction, prevents the method from being used for the routine determination of crystal structure. The solution to that problem — the 'phase problem' of X-ray crystallography — came mainly from the development of mathematical methods that enable phases to be deduced, with the aid of the computer, from X-ray intensity data combined with the fact that there are certain physical boundary conditions such as the finite sizes of atoms and the absence of negative electron density.

The value of the optical analogue continues to be the insight it provides into structure determination, and the special uses to which it can be put, such as the study of disorder and other effects in semi- and poly-crystalline materials. The modern techniques used in producing optical transforms (of which Figs. 5.06 and 5.10 are examples) have largely been developed by Lipson, Taylor and others in Manchester since the 1950s. The *Atlas of optical transforms* (Harburn *et al.*, 1975) contains a wide variety of illustrations.

(iii)   *Lens aperture and aberrations*

In Fig. 5.10(a)(c)(e) circular apertures of diminishing size have been introduced in the centre of the diffraction plane to limit the amount of information allowed to contribute to the formation of the image. By removing the outer diffraction spots the masks increasingly deprive the image of the higher spatial frequencies — an example of **spatial frequency filtering**. The results, shown in (b)(d)(f) are correspondingly deficient in fine detail.

To calculate this effect of the aperture we can argue that it has an aperture function that multiples the diffraction pattern of the object by unity inside its boundary and by zero outside. Since the image is the transform of that part of the pattern in the diffraction plane that contributes to the image, we can then write

$$\text{Image} = T[T(\text{object}) \times (\text{Ap}.\text{fn})] \tag{5.08}$$

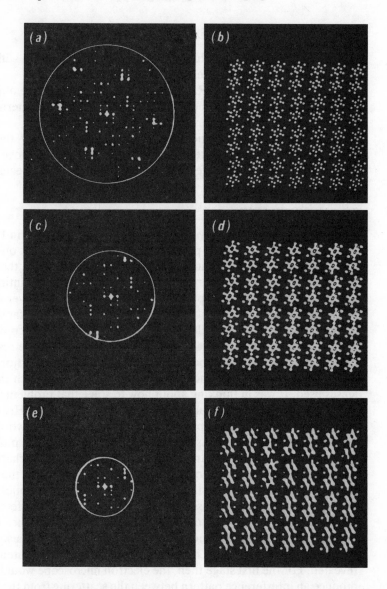

Fig. 5.10 — Deterioration of image quality as high spatial-frequencies are removed by filtering in the diffraction plane. (Reproduced by permission, from Taylor and Lipson (1964), *Optical transforms*. Bell, London.)

where T denotes Fourier transformation, and the appropriate limits must be specified for the aperture function (Ap . fn).

The *equivalence* of this to the Rayleigh convolution treatment of imaging (§5.2) is shown by applying the convolution theorem to the above equation, which immediately gives

$$\text{Image} = \text{Object} \circledast T(\text{Ap}.\text{fn}) \tag{5.09}$$

i.e. the image is the convolution of the object distribution with the Airy pattern given by the lens aperture.

In accordance with the Rayleigh model each point of the object is here being regarded as a source and is being imaged as the Airy pattern of the lens aperture.

With regard to aberrations, their effect is to modify the aperture function. This has no effect on the principles involved above and for a treatment of aberrations reference should be made to the specialist books.

5.4 HOLOGRAPHY

The invention of holography in 1948 by D. Gabor, for which in 1971 he was awarded the Nobel Prize in physics, stemmed from his work on improving the quality of images obtained in electron microscopy. In the 1940s the results obtained with electron microscopes were disappointing because although a hundredfold improvement on the resolving power of the best light-microscope had been obtained, the resolution stopped far short of the theoretical value. The fast electrons used in electron microscopy have a de Broglie wavelength of about $1/20$ Å, so that atoms should be resolved; but the practical limit at that time was approximately 12 Å. A major reason for the shortfall was the presence of aberrations associated with the electron lenses used. It was in thinking about how to solve that problem that Gabor devised the technique he called **wavefront reconstruction**. His inspiration came partly from the principles involved in W. L. Bragg's two-wavelength microscopy (§5.3.3). He reasoned that if he could record the phases as well as the intensities in an electron microscope image, then perhaps he could complete the image formation in an optical system which at the same time could be designed to correct the aberrations in the electron optics. The time-honoured way to record phases was to change them into interference effects and in thinking about how to do that in this particular instance Gabor was influenced by Zernike's successful use of a 'coherent background' to investigate lens aberrations in such a way that phases and intensities were revealed. For the first stage then, the electron microscope would be used to produce an interference pattern between the scattering from the object and the unused 'background' (i.e. the non-scattered part of the illuminating electron beam).

Gabor's original experiments were solely with visible light and used a method that was soon superseded: we need not describe it here. The principle of the method, as opposed to the practical arrangement, is shown in Fig. 5.11. A single object point, B, is considered. Wavefronts scattered from B interfere with wavefronts from a coherent background, or from a coherent reference source such as A in the figure. Suppose that the interference pattern is photographically recorded as a positive which, after processing, will only transmit at the positions of the interference maxima. Then, if the photograph on its own is now illuminated by the reference source, the wavefronts transmitted by it are still in phase with the reference

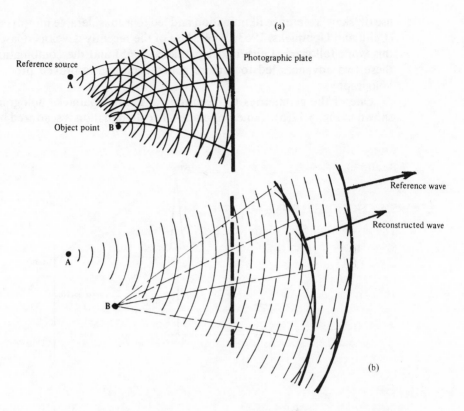

Fig. 5.11 — Gabor's schematic portrayal of wavefront reconstruction: (a) recording,
(b) reconstruction (Gabor, 1972).

source. But as interference maxima they are also in step with wavefronts that
would leave B if B were still there. Thus wavefronts from B could, despite its
absence, be said to have been 'reconstructed'. The photograph has acted as
what Gabor called a **'hologram'** — From the Greek word *holos* meaning
whole — to indicate that it contains the whole of the information about the
object, both amplitude and phase.

Though confirming the general idea, Gabor's results were bedevilled by
the inadequate coherence length (only about 0.1 mm) of the light from the
high-pressure mercury vapour lamp that was used, and by the low level of
the illumination available after the introduction of a small pinhole (3 μm
diameter) to secure adequate spatial coherence. For these and a number of
other reasons, the envisaged application to electron microscopy was unsuc-
cessful. As Gabor described it, holography went into a long hibernation.
The revival came with the work of E. N. Leith and J. Upatnieks (1962).
Their success was due to their recognition of the similarity of Gabor's
wavefront reconstruction process to the theoretical work that Leith and his
collaborators had been undertaking on 'side-looking' radar. It involved the

use of 'skew' reference illumination and led to a considerable improvement (Leith and Upatnieks, 1963, 1964). Use of the recently developed lasers in this work followed (Leith and Upatnieks, 1965) and the combination of these two advances led to a more generalized and improved process of holography.

One of the geometries for recording a 'Leith–Upatnieks hologram' is shown in Fig. 5.12(a). Coherent plane-wave illumination is scattered by (in

Fig. 5.12 — The Leith–Upatnieks method for holography.
(a) Recording. (A prism is commonly depicted in this type of diagram, though a mirror is usually used to deflect the reference beam in practice.)
(b) Reconstruction.

this example) a transparent object, and the hologram is formed by making that scattered illumination interfere with a reference beam produced by suitably deflecting an unused part of the incident illumination. To appreciate how the hologram obtained by photographically recording this interference pattern carries the amplitude and phase data required for 'reconstructing' an image of the object, we need only consider the process in one dimension, the x-axis in the figure.

Let A_x and δ_x be the amplitude and phase, respectively, of the illumination at x, that has arrived from the object. The amplitude, A_0, of the reference illumination at the hologram can be assumed constant, and its phase at any point x is $(2\pi/\lambda)x \sin\theta = 2\pi\alpha x$ say, where θ is the angle of incidence of the reference beam and $\alpha = (\sin\theta)/\lambda$. The total amplitude, R_x, as a function of x, can therefore be expressed as

$$R_x = A_0 e^{i2\pi\alpha x} + A_x e^{i\delta_x} . \tag{5.10}$$

The intensity I_x at x is then given by

$$I_x = |R_x|^2 = A_0^2 + A_x^2 + A_0 A_x e^{-i(2\pi\alpha x - \delta_x)} + A_0 A_x e^{i(2\pi\alpha x - \delta_x)} \tag{5.11}$$

$$= A_0^2 + A_x^2 + 2A_0 A_x \cos(2\pi\alpha x - \delta_x) \tag{5.12}$$

The cosine term represents a fringe system recorded in the hologram. It carries (hence 'carrier fringes') all the information about the object wavefront system because the amplitude, $A_0 A_x$, of the fringes is proportional to A_x (A_0 is constant) and the position of the fringes (spacing $1/\alpha$) with respect to x is determined by δ_x.

Used for reconstruction, the hologram can be illuminated with parallel coherent light at normal incidence (Fig. 5.12(b)). By a suitable photographic development process the amplitude transmission factor in the hologram can be linearly related to I_x in Eqns (5.11) and (5.12). The *amplitude*, $(A_x)_{\text{rec}}$, of the reconstructed wavefront is therefore given directly by the right-hand sides of those equations if we omit a multiplying factor that does not affect the interpretation. Using Eqn (5.11) we have

$$(A_x)_{\text{rec}} = A_0^2 + A_x^2 + A_0 A_x e^{-i(2\pi\alpha x - \delta_x)} + A_0 A_x e^{i(2\pi\alpha x - \delta_x)} . \tag{5.13}$$

The first two terms represent the illumination in the direction of the reconstruction beam. This is essentially A_0^2, as A_x^2 is normally arranged to be much less than A_0^2. The third and fourth terms can be rewritten as follows

$$A_0 e^{-i2\pi\alpha x} A_x e^{i\delta_x} + A_0 e^{i2\pi\alpha x} A_x e^{-i\delta_x}$$

respectively. In the first of these, $A_x e^{i\delta_x}$ is exactly the complex amplitude wavefront distribution that originally left the object to form the hologram. Alone, it would enable a (virtual) image to be formed of the object in its original position. However, multiplication by $A_0 e^{-i2\pi\alpha x}$ imposes a phase shift which amounts to a rotation (Appendix A), making it necessary to view the hologram in direction $-\theta$ (Fig. 5.12(b)) in order to see the image. In the second term, $A_x e^{-i\delta_x}$ corresponds to a (real) image of the complex conjugate of the object, and multiplication by $A_0 e^{i2\pi\alpha x}$ means that it has to be viewed from direction θ. Being the complex conjugate of the object (cf. the negative sign in the exponential) this latter image is inverted in a way that makes it appear to be inside out — technically referred to as *pseudoscopic*.

It will be noticed that in the above simple account of reconstruction, light at normal incidence has been used, not a duplicate of the original reference beam. When the thickness of the recording emulsion is taken into account, however, the conditions do become more critical.

Other arrangements are now commonly used for producing holograms of opaque and transparent objects, for 3-dimensional colour imaging, and for the various applications in microwave technology, acoustics, 'non-coherent photography', non-destructive testing, motion studies, information storage, etc. Descriptions are to be found in most modern mainstream physics textbooks. Many of the useful properties of holograms are not associated, however, with these developments in the techniques for imaging, so we shall not describe them here but will restrict ourselves to some Fourier aspects of holography in §5.5.1(iii).

5.5 OPTICAL PROCESSING

5.5.1 Coherent processing

In describing the two-stage process of coherent optical image formation we have seen that the quality of an image is determined by the information reaching it from the object. Specifically, the quality depends on the faithfulness with which its constituent spatial frequency spectrum is a replica of that of the object. We have also seen that the spatial frequency spectrum is accessible in the diffraction plane. Figure 5.10 showed an example, one in which image quality deteriorates when high frequencies are removed by simply placing an aperture in the diffraction plane to prevent frequencies above a certain value from contributing to the formation of the image — a **low-pass filter.**

That was a very simple example of coherent **spatial (frequency) filtering**, which now we need to consider in more general terms, because there are other types and methods of filtering in the frequency domain whose range of applications is far-reaching.

The basic arrangement is restated in simplified form in Fig. 5.13. The object is in the form of a transparency that transmits a complex amplitude $f(x)$ — keeping to one dimension solely for simplicity in expressing the mathematics. The transparency is illuminated at normal incidence with light from, for example, a laser and beam-expander combination to produce planewave coherent illumination. In the diffraction plane we have the Fourier transform of the transparency,

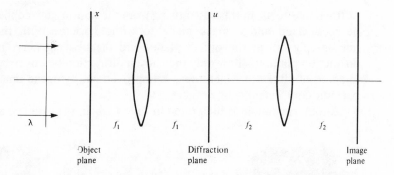

Fig. 5.13 — Basic arrangement for coherent optical filtering.

$$F(u) = \int f(x)\, e^{-2\pi i u x}\, dx \; . \tag{5.14}$$

The image is formed as a further stage of Fourier transformation.

The opportunity for spatially filtering the image is in the diffraction plane and the example (Fig. 5.10) at the beginning of this section is one of **'amplitude filtering'**. (In particular, it is an example of a **'blocking filter'** because of its all-of-nothing nature at any given frequency.)

Filtering with respect to phase, or a combination of amplitude and phase — **complex filtering** — is also practicable.

Optical filtering, then, is a particular form of optical processing in which the spatial Fourier transfrom of an object is operated on in such a way as to have a predetermined effect on the image. It has its origins in the Abbe theory of image formation.

Examples of the three types of filtering will help to give some idea of their range of application.

(i) *Amplitude filters*
A valuable application of this type of filtering was introduced in 1964 by A. Klug and J. E. Berger for the analysis of electron micrographs of large biological molecules by optical diffraction, using the electron micrograph as a diffraction mask.

In many biological specimens there is a component of structural regularity that is of the interest, for example the arrangement of protein sub-units in virus particles. Often, however, the details are not clearly revealed in the electron micrographs because of the presence of, for example, non-organized material in the specimen. The procedure developed by Klug and his collaborators can be described by reference to Fig. 5.13. The electron micrograph is placed in the object plane, and the pattern it gives in the optical diffraction plane is photographed. Only the component in the object that has structural regularity gives a regular diffraction pattern. Other diffraction effects, in this case arising from the non-organized material, can be recognized and a mask made to withold them. With the electron micrograph still in the object plane, and the filtering mask now in the diffraction plane to allow only the chosen diffraction beams to proceed, an image is obtained in which there is improved clarity in the imaging of the periodic component of the object.

Another example is illustrated in Fig. 5.14. In (a) we have an electron

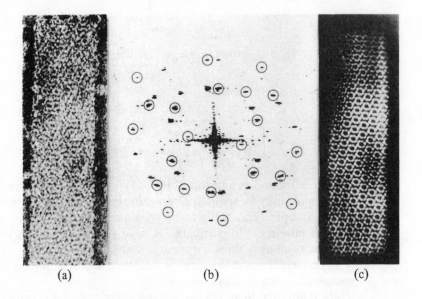

(a)   (b)   (c)

Fig. 5.14 — Optical filtering in electron microscopy.
(a) Electron micrograph of negatively stained flattened tubular structure in a bacteriophage (× 200 000).
(b) Optical diffraction pattern of (a). The circles are round diffraction spots associated with the structure of one side of the flattened tube. A mask is made to allow only these to form the image.
(c) Resulting filtered image, showing helical structure consisting of molecules arranged in hexamers.
(DeRosier, D. J., and Klug, A. (1972), *J. Mol. Biol.*, **65**, 469–488.)

micrograph of a tubular structure found in a form of bacteriophage. The outer coating of the tube is composed of protein molecules of molecular weight approximately 50,000 and arranged in a helix like a coil spring. To obtain this picture the specimen had first been embedded in an electron-dense medium (uranyl formate). This is a standard procedure. The medium scatters electrons more efficiently than the protein, and by occupying the holes and crevices in the surface it throws the surface structure into relief. However, tubular specimens like this become flattened when prepared for electron microscopy, and the details of the back and front of the tube become superimposed. In (b) and (c) we can see how the images can be separated. The optical diffraction pattern of (a) is shown in (b): it was obtained in the same way as for the example described above. Circles have been drawn round the diffraction spots that could be identified as corresponding to the structure of one layer (the back, say) of the tube. With a mask that allowed only these to proceed to the image plane the picture shown in (c) was obtained. The arrangement of the individual molecules (in hexamers) can now be seen clearly. Of course, great care has to be exercised in this work since if an unrelated set of diffraction spots were selected a false image would be produced.

The same principles are now used for processing electron micrographs by computer. The photographic image is converted into digital form by densitometry and a computer is used to perform the Fourier transforms and the filtering. By using this method phases as well as intensities are available and the procedure is generally more amenable than the optical method for correcting for aberrations and other unwanted effects due to the electron optics of the microscope. Regarding an electron micrograph as an aperture function, albeit a very complicated one, its Fourier transform can be calculated in full detail of amplitude and phase. (Because they are not 'seen', phases are all to readily forgotten in optical processing in general; and in work such as the above they can be as important as, or more important than, the amplitudes. As we have noted elsewhere, however, optical methods do have their advantages.)

A further major advance by Klug and his collaborators has been the development of a technique for producing well-resolved 3-dimensional images of large biological molecules (DeRosier and Klug, 1968). Figure 5.15

(a)   (b)

Fig. 5.15 — (a) A 3-dimensional reconstructed image of human wart virus
(c. 500 Å dia.). $(1 \text{Å} = 10^{-10} \text{m})$.
(b) shows the underlying icosahedral surface lattice with the 5-fold and 6-fold vertices marked.
(Klug, A. (1978–9) *Chemica Scripta*, **14**, 245–256.)

shows an example in which a 3-dimensional image of human wart virus has been 'reconstructed'. Calculated by combining the Fourier terms of a number of ordinary 2-dimensional pictures photographed in the electron microscope at different angles of tilt of the specimen, it provides a much clearer image than is obtained in any one of the original 2-dimensionsl views ('projections') alone. This is because it eliminates the confusion present in the individual pictures due to the overlap of features lying within the depth of focus of the electron microscope objective. Prior to this development the latter had been considered to be a limitation inherent in electron microscopy.

For this work Klug was awarded the Nobel Prize in chemistry in 1982.

3-dimensional imaging in this way is an extension of the conceptual framework of X-ray crystallography. The main difference is that in electron microscopy amplitudes and phases can be calculated because, as noted above, an image is available, whereas in X-ray crystallography only X-ray diffraction intensities can be measured.

The same principles underlie the new techniques of computed tomography (Chapter 6).

Insofar as we have mentioned the importance of *phase* and the need for phase data, we have trespassed somewhat on the following paragraphs. Other applications of *amplitude* filtering include the removal of raster from TV pictures, removal of the dot structure of half-tone pictures (this is the 2-dimensional equivalent of raster removal) removal of additive noise, and image contrast control. The last of these is done by altering the balance of high- and low-frequency contributions to images.

(ii) Phase filters

We have referred in (i) to the importance of, and the frequent need for, phase data. An application of filtering in which phases play the leading role has been in the field of optical microscopy.

Sections of biological materials examined under an optical microscope are often almost, or totally, transparent. To say the least, this makes it difficult to observe their structure, unless some trick is used. With such materials it is their refractive index, and therefore their optical thickness, that varies from point to point. Because this only introduces phase differences between light that has passed through different regions it has no effect on the amplitude of the transmitted light and so is invisible to the human eye. For obvious reasons, materials of this type are referred to as **phase objects** as opposed to **amplitude objects** with which we have been mainly concerned.

The classical way of revealing the structure of materials like this is to use selective chemical staining, and the value of that method cannot be overemphasized. A more direct method, however, for distinguishing between regions of different refractive index (and therefore of different biological composition) is by **phase-contrast microscopy**, originally proposed by F. Zernike in c. 1935 (see Zernike, 1942), for which he was awarded the Nobel Prize in 1953 for physics.

Figure 5.16(a) illustrates, in a simple way, the basic principle involved. The phasors labelled A represent the illumination at two points in the image

Fig. 5.16 — Phase-contrast microscopy.
D, diagram; C, condenser; S, specimen; O, components of objective;
P, phase plate; E, eyepiece.
(After Barer, 1955.)

plane of a phase object. They have the same amplitude, A, but their phases differ because of differences in the optical thickness of the object. If illumination of constant amplitude and phase is superimposed across the field, as represented by the dotted phasors labelled B, interference occurs and the resultants, R, are different in amplitude as well as phase. The optical path differences have therefore been made 'visible'. One of the simplest arrangements for doing this is shown schematically in Fig. 5.16(b). Light is focused on the specimen S by a condenser lens C. A diaphragm D in the form of an annular slot is placed in the focal plane of the condenser, and its image is formed as a ring of light in the back focal plane (P) of the objective O by light that has not been diffracted by the specimen. The effects illustrated in (a) be achieved, therefore, by introducing a phase difference between this 'reference' illumination and the light diffracted by the specimen. (The small proportion of the *diffracted* light that passes through the ring in the back focal plane simply dilutes the effect.) The phase difference is introduced by means of a **phase plate** at P consisting of an optically parallel glass plate on

(a)   (b)

Fig. 5.17 — An example of phase-contrast microscopy. Optical micrographs of squamous cells from the lining of the mouth. (Field of view c. 100 µm.) (a) Phase contrast. (b) Conventional bright-field illumination.
(Courtesy of Dr Martin C. Steward, Medical School, University of Manchester.)

which a thin layer of dielectric material has been deposited over the area of the ring, or everywhere except the ring, giving 'negative' or 'positive' phase contrast, respectively. The example shown in Fig. 5.17 is of squamous cells from the lining of the mouth. Details of the cell architecture are more clearly seen with phase contrast (a) than with normal bright-field illumination (b).

More complicated methods of what can collectively be called **interference microscopy** have been developed since Zernike's pioneering work. They generally involve some form of beam splitter and the controlled introduction of a phase difference in one beam before recombination. The Nomarski differential–interference–contrast microscope is particularly popular at the moment and is well described and referenced by Padawar (1968).

(iii) *Complex (holographic) filters*
With its ability to record phase, holography has led to considerable advances in the use of filtering.

The idea of holographic filters was first seriously mooted by A. Vander Lugt in 1963 (his 1964 paper is more accessible) in connection with their possible use in signal detection. Since then their use has been extended to include correction ('balancing') of aberrations in optical systems, image-motion compensation, etc. Before considering applications we need to know the basis on which this type of filtering operates.

The type of hologram we have already considered is described as either a Fresnel or a Fraunhofer hologram depending on whether the photographic plate used for recording it is located in the near- or far -field. Such holograms are used in 'lensless' photography of 3-dimensional objects and for many other applications of holography. A hologram can be formed in any plane,

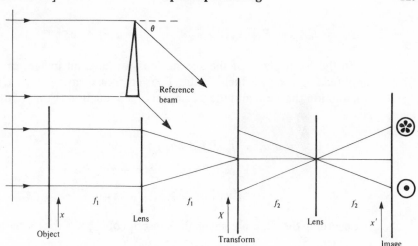

Fig. 5.18 — Coherent optical processing. Optical correlator using a matched spatial-filter for pattern recognition.

however, and with the arrangement in Fig. 5.18 it is recorded in the focal plane of the first lens, the Fourier transform plane of the object. This **Fourier transform (or generalized) hologram** has properties of special value in certain types of filtering. (There is also a 'lensless' geometry for recording this type of hologram.)

Restricting this treatment to one dimension for clarity, the x-axis as indicated in Fig. 5.18, let f_x be the complex amplitude transmitted by the coherently illuminated object. In the focal plane of the first lens we have the Fourier transform of f_x, which we can denote by F_X. The capital F represents the transform of f in the conventional way and X denotes the coordinate in the Fourier plane. (To allow us to see the relationships as clearly as possible we omit various factors such as the wavelength of the illumination, focal lengths of lenses, etc., which only affect the geometrical aspects.)

Let $a_0 e^{-2i\pi\alpha X}$ be the complex amplitude of the reference illumination arriving at the transform plane, where $\alpha = (\sin\theta)/\lambda$ as in §5.4.

The intensity distribution, I_X, in the transform plane is given by

$$I_X = |a_0 e^{-2\pi i\alpha X} + F_X|^2$$
$$= a_0^2 + |F_X|^2 + a_0 F_X e^{2\pi i\alpha X} + a_0 F_X^* e^{-2\pi i\alpha X} . \qquad (5.15)$$

This is photographically recorded as a transparency as described earlier, so that the hologram has an amplitude transmittance proportional to this intensity distribution.

Now consider what happens when this hologram is put back in the transform plane, the reference illumination shut off, and a different object used, transmitting with a different pattern of complex amplitude, a_x say. In the Fourier plane this latter is A_X. 'Filtered' in the Fourier plane by the hologram whose complex amplitude transmittance is given by Eqn (5.15), the amplitude leaving the transform plane, U_X say, is therefore

$$U_X = a_0^2 A_X + |F_X|^2 A_X + a_0 F_X A_X e^{2\pi i a X} + a_0 F_X^* A_X e^{-2\pi i a X} . \tag{5.16}$$

In the focal plane of the second lens we have an image, of complex amplitude $\psi_{x'}$ say, which is the Fourier transform of U_X. Using the convolution theorem and the transform of a δ function (Eqn (4.20)) freely, gives

$$\begin{aligned} \psi_{x'} = {}& a_0^2 a_{x'} + (f_{x'} \circledast f_{x'} \circledast a_{x'}) \\ & + [(a_0 f_{x'}) \circledast a_{x'} \circledast \delta(x' + a)] \\ & + [(a_0 f_{-x'}^*) \circledast a_{x'} \circledast \delta(x' - a)] . \end{aligned} \tag{5.17}$$

(Concerning the negative sign in the subscript of $f_{-x'}^*$ see the footnote to Eqn (4.51) (p. 95).)

The first two terms refer to illumination centred on the axis in the image plane and is not normally of any use. The third term is the convolution of $f_{x'}$ and $a_{x'}$, and is shifted by the δ function to be centred at $x' = -a$: we will call this the **convolution image**. The last term is likewise the convolution of $f_{-x'}^*$ and $a_{x'}$ and is centred at $x' = a$. It is also the cross-correlation of $f_{+x'}^*$ and $a_{x'}$ (see Eqn (4.46)) and is referred to as the **cross-correlation image**.

Note that the above findings are an example of how multiplication in the Fourier (diffraction) plane corresponds to convolution in the image plane.

The hologram of f_x above is referred to as a **Vander Lugt filter** and the following examples will illustrate the uses of the two displaced images.

THE CROSS-CORRELATION IMAGE
Example: pattern recognition (character recognition).

Suppose that one wishes to identify a particular 'pattern' in a series of 'object' transparencies. A Fourier transform hologram filter of the particular pattern, such as our f_x above, is made in the manner we have described and replaced in the transform plane. The identity parade of object patterns being inspected are located in turn in the object plane in Fig. 5.18. When it is the turn of the pattern being sought the cross-correlation image will have a bright central spot. All patterns which do not match f_x will give zero illumination in that image (cf. §4.7) — unless there is some similarity, when a weak 'parasitic response' will be obtained there.

Filters of this type, comprising the Fourier transform of a sought pattern, are collectively known as '**matched filters**'.

THE CONVOLUTION IMAGE
Example: deblurring, aberration compensation.

From earlier chapters it will be evident that the effect of aberrations, or any other feature of an imaging system that results in a blurring of the image, can be regarded as a convolution of the ideal image, a_x say, with a blurring function, g_x (cf. §5.3.3(iii)). The normally produced optical image would therefore be the convolution $a_x \circledast g_x$. To remove the effect of g_x one can therefore multiply (*vide* the convolution theorem) the transform of this convolution $(A_X G_X)$ at the diffraction stage, by $1/G_X$. To make a filter to do this, light from a point source is passed through the aberrated lens system and interfered with a collimated reference beam. There are snags, however, and the reader should consult the specialist books for details.

•   •   •   •   •

A disadvantage of processing under coherent conditions is that the handling of complex amplitudes precludes the use of TV displays or LED arrays as inputs to the processing system. Furthermore, coherent processing tends to be troubled by 'noise' produced by dust, scratches and other blemishes on optical components. Also, when the output from such systems is presented it is usually in the form of intensities, so that phase data are lost.

5.5.2 Incoherent processing

Because of the disadvantages inherent in coherent processing mentioned in the previous section, a trend during the past decade has been towards incoherent processing. Needless to say, there are disadvantages in that too and choice must depend on the nature of the problem to be handled. With incoherent processing, inputs and outputs normally need to be non-negative and real. Bipolar or complex-valued processing can be performed, however, by multiplexing — but then some form of hybrid system is usually required.

Three main kinds of incoherent processing can be identified and these are briefly described below but only in the simplest outline because they refer to topics that are really outside the subject of this book. The first is again concerned with the use of diffraction, the second is based on purely geometrical principles in which diffraction plays no active role, and the third is concerned with computer-based digital processing.

(i) *Diffraction-based processing*

Example: pattern recognition by energy-spectrum correlation.

With the arrangement shown in Fig. 5.19 a transparency T_1 is illuminated with coherent plane-wave quasimonochromatic light. Its transmitted complex amplitude can be denoted by f_x and its Fourier transform in the focal plane of lens L_1 by F_X, using 1-dimensional representation again for simplicity. A diffuser in the transform plane destroys the coherence there and produces an incoherent, effectively 'self-luminous', intensity distribution proportional to $|F_X|^2$ which is the energy spectrum of f_x (cf. §4.7.1).

Another transparency is located at T_2. The output from the system is therefore the convolution of $|F_X|^2$ with the energy spectrum of the transparency at T_2. Since T_1 and T_2 are normally real, if either is inverted the convolution is equivalent to correlation. The instrumentation represented

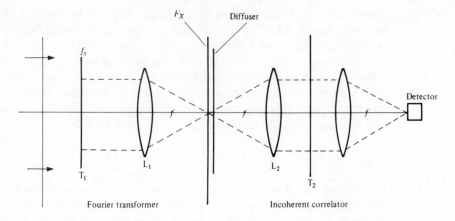

Fig. 5.19 — Incoherent optical processing. Energy-spectrum correlator.

schematically in the figure is commonly known as an **energy-spectrum correlator**.

Used for alphanumeric pattern recognition, for example, the technique depends on the fact that only if the two transparencies T_1 and T_2 have the same spatial frequency spectra is the cross-correlation output from the system (usually measured with a photo-detector on the axis) significantly non-zero. A transparency of the 'required' pattern is used for T_2 and the unknowns to be searched are located at T_1. Success depends upon the various spectra all being different. Against this shortcoming the method has a distinct advantage over that described in §5.5.1 in that the measured cross-correlation output is independent of the position and orientation of the patterns in the input plane because their spectra are not directional.

Applications include alphanumeric character recognition in language translation, information retrieval, etc.

(ii) *Geometrical optics-based processing*
Example: pattern recognition.

In Fig. 5.20 light from a source S is collimated by lens L_1 and passes through a photographic transparency T_1 and then through a second transparency T_2 that can be displaced in its own plane. Lens L_2 focuses the light transmitted through T_1 and T_2 on to a photo-detector at its focal point.

Let the intensity transmittance of T_1 and T_2 be $t_1(x, y)$ and $t_2(x + u, y + v)$ respectively, where u, v is the translation of T_2 with respect to T_1.

The detector integrates the light transmitted by the two transparencies, and its output, P, a function of u, v is given by

$$P(u, v) = \iint t_1(x, y)\, t_2(x + u, y + v)\, dx\, dy \ . \qquad (5.18)$$

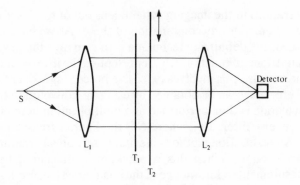

Fig. 5.20 — Geometrical optics-based processing. T_1, T_2 transparencies in x, y plane perpendicular to the figure.

This is the cross-correlation of $t_1(x)$ and $t_2(x)$ (cf. Eqn (4.40)) and from §.4.7 it will be obvious that in general a strong signal is detected only if the two patterns comprising f and g are the same and if $u = v = 0$.

This **optical multiplier and integrator** can therefore be used for pattern recognition though, unlike the energy-spectrum correlator in (i), it requires that for a pattern T_2 to be identified with a pattern T_1 the two must both be in the correct rotational orientation.

This brief note is only intended as a hint at the types of processing that are possible with geometry-based incoherent imaging and related systems. They include spatial and temporal scanning methods, multichannel processing, etc. and are designed around a wide variety of devices (LEDs, charge-coupled device (CCD) arrays, etc.).

(iii)   *Digital computer-based processing*
F. H. Preston
Example: deblurring

Digital as well as optical methods can be used to remove or reduce the effect of a blurring function. In practice there are a number of problems but although a perfect restoration cannot be realized usable improvements can be made.

The image first has to be acquired or converted into digital form. A digital picture is made up of an array of individual picture elements — **pixels** — each having an appropriate brightness and colour depending on its position. Each pixel is represented by a binary code in the computer and an image frame store and television monitor can be used to display the picture. Digitization is normally achieved by scanning the image with a focused beam of light, measuring the intensity of the transmitted (or reflected) light, and converting it into digital form.

Blurring in an image can arise for a number of reasons including aberrations in the imaging system, lens out of focus, subject and/or camera movements, or any combination of these. As we have noted in §5.5.1, the basis of the deblurring technique is to convolve the degraded image with a spatial domain function, $1/g$, representing the inverse of the blur function, g. And because this involves extensive processing, advantages in computational efficiency can be obtained by using the convolution theorem (§4.5). The calculation is then performed as a multiplication in the Fourier domain, by the inverse filter, $1/G$, where G is the Fourier transform of g.

All restoration methods inevitably rely on an accurate estimation of the blur function. When this is not known, or cannot be determined from measurement or knowledge of the imaging system used, it has to be obtained from the blurred image itself. Cannon (1976) has shown that a fruitful approach is to use the information present in the average power spectrum (§4.7.1) of the blurred image. By dividing the image into subsections, determining the squared modulus of the Fourier transform of each and taking the average, the contributions from the spatial structure common to all the subsections is emphasized relative to overall picture information. This method works well in the case of simple blurs like those caused by camera shake where the whole image is degraded by the same blurring function. However, there are cases where the blur changes with position in the image, for instance subject motion against stationary surroundings or a focus blur varying from foreground to background. To treat these cases, appropriate software can be written allowing the size of the subsections and the area over which they are averaged to be chosen by the operator. However, the Cannon method relies on the subsections being significantly larger than the blur and the blur being non-varying over the area averaged. Where this cannot be achieved, or for complicated blurs, the alternative is to guess, albeit using clues in the blurred image. Streaks which can be associated with point or near-point highlights and features which can be recognized as being sharp in the original scene are good indicators.

Fig. 5.21(a) shows an average power spectrum for the case of the simple linear motion blur involved in photographing a road sign (b) from a moving car. The spectrum was obtained by dividing the picture into squares and computing the average of the squared modulus of the 2-dimensional Fourier transform of each. It contains dark bands of a spatial frequency and direction dependent on the blur. Calculation of the Fourier transform of this (actually calculated as the Fourier transform of the logarithm of the power spectrum — the average power '**cepstrum**') allows these features to be easily related to the blur. In the case of the simple linear motion blur in our illustration here, this results in two spots as shown in (c). The separation of the spots is twice the length of, and in the same direction as, the blur function (g); allowing the latter to be deduced as shown in (d).

Having obtained an estimate of the blur function, g, multiplication by the straight inverse filter, $1/G$, seems attractive because of its simplicity but it has two main drawbacks. Real blurs often produce zeros in G leading to singularities in the inverse filter. Furthermore the simple inverse filter also tends to amplify the high spatial frequencies, greatly accentuating any noise

Fig. 5.21 — Digital deblurring.
 (a) Average power spectrum of (b).
 (b) Road sign photographed from a moving car.
 (c) Fourier transform of the logarithm of (a) (the power 'cepstrum').
 (d) Blur function deduced from (c).
 (e) Deblurred image.
(a), (c) and (d) are enlarged twice actual size for clarity.
(Courtesy of Dr F. H. Preston, Scientific Research and Development Branch, Home
Office, St Albans, Hertfordshire, UK.)

in the original. To overcome these problems a number of variations on the inverse filter have been proposed (Pratt, 1978) and can be used successfully to restore blurred images (Lehar *et al.*, 1982). Their advantage is that the restoration is achieved in a single convolution. The restoration shown in (e) was performed in such a way.

Another method is to use an iterative restoration technique. One such technique which has been used with good success is the maximum entropy method of Gull and Daniell (1978) and Skilling (1981) (Appendix H). It is also being extended to blind deconvolution where the algorithm does not require the operator to estimate the blur function since this is determined solely from the blurred image.

●   ●   ●   ●   ●

Articles on applications and developments in the field of data processing are to be found in volumes specially dedicated to this field: they appear, for example, in the series published by Springer-Verlag under the general titles *Topics in Applied Physics* and *Topics in Current Physics*, and in the series *Progress in Optics* edited by E. Wolf and published by North-Holland, Amsterdam.

6

Image reconstruction from projections (medical imaging)

6.1 INTRODUCTION

In Chapter 5 we have briefly noted that in the electron microscopy of large biological molecules, pictures taken from different directions can be used to reconstruct, mathematically, an image that is clearer than any of the original photographic 'projections' alone (Fig. 5.15). This is an example of one type of 'image reconstruction from projections'.

As long ago as 1917 the Austrian mathematician J. Radon, working on gravitational theory, showed theoretically that an object can be reconstructed from the infinite set of all its projections. The first reconstructed images were, however, obtained independently of Radon's work, in radioastronomy (Bracewell, 1956). The application in electron microscopy mentioned above was developed, again independently, by DeRosier and Klug in the late 1960s.

In this chapter we examine the principles underlying this type of imaging, mainly in the context of medical applications, where it has led to very important advances in the techniques available for medical research and clinical practice in recent years.

In medicine there are needs for information about the distribution of various properties, chemical and anatomical, in the human body. Obviously, it is preferable if those data can be obtained by non-invasive methods. The conventional **X-radiograph**, such as a 'chest X-ray' is an example that immediately comes to mind, and at the same time it illustrates how there was a need for the advances that we are concerned with here. For one disadvantage in the conventional technique is its inability to discriminate between structures that overlap in the direction of viewing. The new method of image reconstruction, that of X-ray **computed tomography** (from the Greek, and loosely translating to mean section drawing) has overcome this (§6.2). X-ray measurements are made from many different directions and combined mathematically to reconstruct cross-sectional maps, X-ray **tomograms**, of any part of the body, with unprecedented fineness of detail and lack of masking by structures in adjacent planes. Alternative terms for computed tomography (CT) are **reconstruction tomography** and **computer-assisted tomography** (CAT) and the machines are generally called **scanners**.

X-ray imaging is mainly concerned with the anatomical structure of the body, distinguishing between different components that have sufficiently different absorbing powers (different '**densities**') to X-rays. Certainly the latest scanners can distinguish between brain tissue, fatty tissue, cerebrospinal fluid etc., and the combined use of injected contrast liquids and reconstruction techniques has proved to be a powerful diagnostic tool for special cases (Fig. 6.06(a)(b)). However, X-ray tomography tells us little about the functional or physiological state of the internal organs of the body. Moreover, diseased tissues usually have X-ray absorption properties so similar to those of normal tissue that pathological lesions (e.g. tumours) can go undetected in CT scans unless they are large enough to change the shape and/or size of the affected organ.

Other modalities of computed tomography, e.g. **nuclear magnetic resonance** and **emission computed tomography**, have been developed for their greater sensitivity in meeting some imaging needs, and for the study of the chemistry and physiology of the body. These are described briefly in the later sections of this chapter. Although, as physical probes, they are very different from X-rays, the optical principles involved in the actual image reconstruction techniques are similar and will, therefore, not be considered in detail in this book.

In the rather complex evolution of all these new methods of medical imaging the major contributions that stand out include those of Cormack who became interested as early as 1956 in corrections for body inhomogeneities when calculating radiotherapy dosages (see especially Cormack, 1963); Oldendorf (1961) who first recognized the potential of reconstruction tomography as a clinical imaging technique; Kuhl and Edwards (1963, 1968) who introduced imaging by **isotope scanning**, subsequently to be developed into the technique now referred to as **emission computed tomography** (§6.3.1); Hounsfield with the EMI Patent filed in 1968 and production in 1971 of the first commercial X-ray scanner, the EMI Mk I head scanner (Hounsfield, 1973), soon followed by the whole-body scanner (Ledley *et al.*, 1974); Lauterbur who in 1973 triggered the application of **nuclear magnetic resonance** (NMR) for medical imaging (§6.3.2). In addition there are important developments in computed tomography using ultrasound (§6.3.3).

The 1979 Nobel Prize in physiology/medicine was awarded jointly to

Cormack and Hounsfield for their pioneering work in reconstruction tomography.

6.2 X-RAY COMPUTED TOMOGRAPHY

6.2.1 Introduction

With the traditional method of X-radiography still widely used, and adequate for many purposes, a projected image (or **shadowgram**) of the object is obtained, in which structural details are revealed by point to point variations in '**density**', i.e. variations in the absorptivity to X-rays, of the body. Inevitably, there are difficulties: different structural components may not be of sufficiently different density to make them distinguishable, and/or their images may overlap, totally or partially, in the viewing direction. Digital methods introduced in recent years have helped to overcome such problems. These methods have been made possible by the development of large-area X-ray detectors for signal conversion, and computers capable of high speed digital data processing. (An example of the use of this **digital radiography** is the investigation of blood vessels such as arteries. The time-honoured procedure is to increase the X-ray density of the blood by injecting a contrast medium — usually a compound containing iodine — but even then delineation of the blood vessels is still difficult by ordinary inspection of the radiographs because the concentration of the medium in the blood has to be kept quite low. Digital subtraction of X-ray images recorded before and after injection, however, yields much clearer pictures. Another example of the use of digital subtraction is in the elimination of bone shadows from chest radiographs. Two images are obtained with different X-ray energies; one easily transmitted by bone and the other appreciably absorbed by bone, but both readily transmitted by the surrounding relatively low-density tissues or organs.)

In many instances, however, the overlapping of images in a single viewing direction does pose serious problems, and even the inspection of '**projections**' taken from different directions is no real solution because each projection poses its own overlap problems. In modern computed tomography the experimental technique itself is different, in that the X-rays only pass through the actual layer ('**slice**' or '**section**') whose structure is to be imaged, thus avoiding the inclusion of unwanted information from other regions of the body. (In an earlier form of tomography, described as long ago as 1915 by Baese and used in certain types of investigation since the 1930s, the X-ray beam travels through the body in the direction of viewing as in normal radiography, but coupled, parallel (but opposite) movement of X-ray tube and photographic plate during exposure causes blurring of all but one chosen plane through the body. Now obsolete, the details of this '**focal-plane**' (or '**focused**') **tomography** are irrelevant here.)

The simplest arrangement used in modern computed tomography is shown in Fig. 6.01(a)(b). It is known as the **translate–rotate** method. A 'pencil' of X-rays scans the chosen layer from a number of different

Fig. 6.01 — X-ray computed tomography.
    (a) For the imaging of a given section all measurements are made in the
        plane of the section. The particular X-ray source — detector
        movement indicated here is for the 'translate–rotate' method (b).
    (b) Parallel-ray geometry: plan view of the translate–rotate method.
        For each angle of viewing, a set of ray integrals yields the 'pro-
        jection' for that direction. A large number of viewing directions is
        needed, of which only three are shown here.
    (c) Fan-beam geometry: plan view. The alignment of the X-ray tube
        relative to the detector array is maintained as both rotate in the
        plane being imaged.

directions, but always in the plane of the layer. For each direction, the
absorption of the X-rays is measured along an evenly spaced set of '**ray-
paths**'. The measurement for each individual ray-path gives a '**ray integral**'
and the set of these ray integrals, obtained by stepwise parallel *translation* of
the X-ray tube and detector, gives the 'projection' of the structure of the
plane in the given direction. The tube–detector assembly is then *rotated*, still
in the plane being examined, through successive angular steps, to enable a
large number of projections to be obtained.

    To increase efficiency, the actual methods used in practice are rather
more complicated than this (e.g. Fig. 6.01(c)), though in some commercial
units the measured X-ray data are all converted to be equivalent to the
parallel-ray geometry we have described, for ease of computation. We
mention this again later (§6.2.4) since, at the moment, the need is to see how
projections actually provide the information required for image reconstruc-

tion. The problem is rather like trying to obtain an aerial map of a forest by taking photographs at ground level through the forest, and only from the edges. As in all survey work, we first need to familiarize ourselves with the coordinate system used.

Any point in the plane of a section to be reconstructed can be defined in the normal way by x, y Cartesian coordinates (Fig. 6.02), but tomographic

Fig. 6.02 — 'Projections' in computed tomography.
　　　　　The 'density' distribution in a section is specified with respect to the fixed axes x, y. X-ray absorption measurements are made in many directions in the plane of the section and are referred to axes r, s which rotate so that the X-ray measurements are made along s, at ϕ to y. The projection for a given ϕ, along r, is expressed as $[p(r, \phi)]_\phi$.

projections are described by an orthogonal r, s coordinate system which is rotated through a sequence of angles, ϕ, relative to the x, y axes, so that the s axis can coincide successively with each chosen projection direction. The individual ray paths comprising a projection are then specified by r, ϕ where r is the perpendicular distance of each particular ray from the chosen origin, and ϕ is the angle between the projection direction and the y-axis. If a ray passes through the point x, y at an angle ϕ, and its distance from the origin is r, we have in general, from Fig. 6.02:

$$r = x \cos \phi + y \sin \phi \ . \tag{6.01}$$

Let us consider the passage of X-rays along one ray path. An X-ray beam loses intensity as it passes through matter in accordance with the **exponential law of absorption**†. The intensity, I_0 say, of incident X-rays of a given wavelength is reduced after passing through a thickness t of a homogeneous material of '**linear absorption coefficient**' (μ) for that wavelength, to a value I_t given by

$$I_t = I_0 \, e^{-\mu t} . \tag{6.02}$$

(With X-rays one should strictly refer to **attenuation** rather than absorption because various processes other than absorption are involved. However, these are normally all combined for the purpose of defining μ.)

For a given ray path through the inhomogeneous type of body, human of otherwise, that is the usual subject of radiological examination, the 'density' (in terms of μ) varies from point to point within any section: it will be denoted here as $f(x, y)$, the **density function**. Along a ray path in the s direction, for a given r, the μ in Eqn (6.02) becomes an integral as follows

$$I = I_0 \exp\left[-\int_s f(x, y) \, ds \right] . \tag{6.03}$$

Mathematically, the integral is a **line integral**, though — as mentioned earlier — it is referred to in this work as a **ray integral**, the ray concerned being defined by r and ϕ. For a given ϕ, the value of the ray integral (with

---

† The **exponential law of absorption** states that the fraction dI/I of the intensity lost when radiation of intensity I, of a given wavelength, traverses an infinitesimal thickness dx, is proportional to dx for all values of I. This is expressed as

$$\frac{dI}{I} = -\mu \, dx$$

where μ is a constant for the given wavelength, and the minus sign denotes reduction of intensity with increasing x. For a finite thickness, t, we integrate the above. Thus we have

$$\int_0^t \frac{dI}{I} = -\mu \int_0^t dx$$

and thence $\log_e\left(\dfrac{I_t}{I_0}\right) = -\mu t$

i.e. $I_t = I_0 \, e^{-\mu t} .$

as in Eqn (6.02). In optics, this is **Lambert's law**.

respect to s) expressed as a function of r, is then the '**projection**' of the section and is denoted as $p(r, \phi)$ (Fig. 6.02). It is with that meaning of $p(r, \phi)$ that workers in this field write

$$p(r, \phi) = \int_s f(x, y)\, ds \ . \tag{6.04}$$

Substituting for the integral with the aid of Eqn (6.03) we have

$$p(r, \phi) = -\log_e (I/I_0) \ . \tag{6.05}$$

I/I_0 can be measured with an X-ray detector, and $p(r, \phi)$ can, therefore, be readily determined experimentally as a function of r, for each chosen value of ϕ.

We are now ready to move on to the central problem in X-ray computed tomography, namely how to derive $f(x, y)$ from projections obtained at different ϕ, denoted here as $p(r, \phi)$. As a start, it is sensible to look at the most obvious method, that of simple back-projection (§6.2.2), because although it is unsatisfactory and not now used as such, it lies at the heart of, and is therefore essential for an understanding of, methods that have successfully been put into practice. There have been two main areas of development. The first, based on iterative reconstruction (§6.2.3), is a modification of the simple back-projection method, in which the deficiency of the back-projection method is to some extent overcome by what amounts to successive approximation. In the second (§6.2.4) the Fourier relationships between an object and its projections are realized. These are the most powerful methods and have been made usable as a result of the developments in computer hardware and software.

6.2.2 Reconstruction by simple back-projection

The obvious way to attempt to reconstruct an image from projections at different angles is to project each one backwards, as shown schematically in Fig. 6.03. It can be done mathematically or by an analogue method (several of the latter were devised originally but are not used now). As the actual density distribution is unknown, one has no choice but to assume that the experimentally measured absorption along a ray path is due to a uniform distribution of density along its length. In the simple example in Fig. 6.03 this procedure results in the image consisting of a star-shaped distribution instead of the small circular patch that actually gave the measured projections. The star shape can be described as the **point-spread function** of the method (cf. §2.3). With a very large number of projections the star effect can be minimized, but then there tends to be an unacceptable level of general blurring of detail in the image.

(Since the density at any point in the image reconstructed in this way is the sum of all the ray projections that pass through that point, the method is

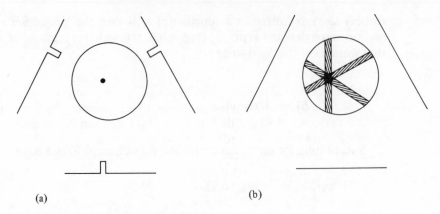

Fig. 6.03 — Reconstruction by simple back-projection.
(a) Three projections of a circular section that contains one small region of high density.
(b) Image reconstruction by back-projection, from the three projections in (a). For each projection the actual distribution of density along the ray paths is unknown. It therefore has to be assumed that the density distribution is uniform along each ray path. The small region of high density actually present in this example is consequently incorrectly reconstructed as a 'star'.

also referred to as the **summation method** or the **method of linear superposition**.)

The reconstruction is expressed mathematically as

$$f'(x,y) = \sum_{j=1}^{m} p(r_j, \phi_j) \Delta\phi$$

i.e.   (6.06)

$$f'(x,y) = \sum_{j=1}^{m} p(x\cos\phi_j + y\sin\phi_j, \phi_j) \Delta\phi \;,$$

using Eqn (6.01) for r, and where m is the number of summations and $\Delta\phi$ is the angular separation between adjacent projections, i.e. $\Delta\phi = \pi/m$. Note that in using this expression to calculate the value of $f'(x,y)$ at some particular point, x_1, y_1 say, only the value of the one ray integral through x_1, y_1 in each projection is used. That is why the summation is only for the number of projections, m.

This was the method used by Oldendorf in his pioneering work in the early 1960s, though with a different scanning arrangement from that employed today. One way of overcoming the shortcomings that we have noted would be to find a way of avoiding the utterly unjustifiable assumption that there is a uniform distribution of density along each ray path. That problem was tackled with some success and is the subject of the next section.

6.2.3 Iterative reconstruction

This method is based on the solution of Eqn 6.04 (with $p(r, \phi)$ determined via Eqn (6.05)) for the density function, $f(x,y)$, by the mathematical procedure of **iteration**. It is a method of successive approximation in which corrections are sequentially applied to an arbitray starting solution, to bring that solution progressively into a satisfactory agreement with the experimental data. We shall see that it is a logical extension of the simple back-projection method described in the previous section.

First, it is necessary to recognize that the integration in Eqn (6.04) is, in practice, a summation because (a) measurements are made at discrete, spatial intervals, and (b) the pencil-beam of X-rays is of finite cross-section. A layer being imaged has to be thought of as consisting of an array of cells or picture elements ('**pixels**') of finite size (Fig. 6.04) (or '**voxels**' if one includes

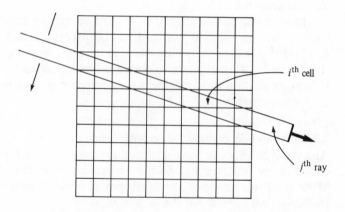

Fig. 6.04 — The array of 'pixels' comprising a section to be imaged. The contribution of the ith cell to the jth 'ray' (really a strip of finite width) is referred to as w_{ij}.

their thickness). And instead of 'rays' one is really working with strips, though the former term is usually retained.

For the jth ray in Fig. 6.04 the ray integral can be expressed as the '**ray sum**',

$$p_j = w_{1j} f_1 + w_{2j} f_2 + \ldots$$

i.e.

$$p_j = \sum_i w_{ij} f_i , \tag{6.07}$$

where w_{ij} are 'weighting factors' that allow for the fact that each cell does not necessarily contribute to the absorption along a given strip.

To solve this equation for f_i by iteration (the number of cells is normally too large to permit its solution as a set of matrix equations) a set of f_i values is first chosen arbitrarily, even putting $f_i = $ a constant. Ray sums, p_j, are then calculated using the known w_{ij}, and each calculated value is compared with the corresponding value determined experimentally via Eqn (6.05). If a calculated ray sum is, for example, smaller than its measured value, then the f_i of every cell along that particular ray path is scaled up by the required amount. Done for all the cells involved in all the ray sums, the first iteration is complete. So far, this amounts to no more than simple back-projection because it assumes that there is uniform distribution of density along ray paths. However, since most cells are involved in more than one ray sum (the ray paths cross in many different directions) each correction is likely to affect the agreement for some of the other ray sums. Further iterative cycles are therefore performed and these normally converge to give an acceptable level of agreement between calculated and experimental data. (Several methods of applying the corrections have been devised but the details need not bother us here.)

Iterative reconstruction proved to be a valuable advance on the simple back-projection method outlined in the preceding section. As already mentioned, it was used by Hounsfield in the first commercial X-ray scanner (Hounsfield, 1973). It was also employed, independently, by Bracewell in 1956 in connection with imaging in radioastronomy, and by Gordon *et al.* in electron microscopy (1970).

6.2.4   Analytical methods of reconstruction
Analytical methods of reconstruction are based on the Fourier coefficients of the image being related to the Fourier coefficients of the projections. Made possible, mainly by the development of fast digital computers, these methods have superseded the earlier ones.

(i)   *Fourier reconstruction*
The actual **density function**, introduced as $f(x, y)$ in §6.2.1, can be expressed as a Fourier integral, or Fourier series in the unlikely event of $f(x, y)$ being periodic. As a logical extension of the 1-dimensional Fourier integral in Chapter 4 (Eqn (4.08)) we can write

$$f(x, y) = \int\limits_{-\infty}^{+\infty}\!\!\int F(u, \upsilon)\, e^{2\pi i(ux + \upsilon y)}\, du\, d\upsilon \ . \tag{6.08}$$

This describes $f(x, y)$ as a superposition of sinusoidal density 'waves' crossing in various directions in the x, y plane. The real and imaginary parts in the exponential notation represent the cosine and sine contributions in the usual way (Appendix A), and u, υ are the spatial frequencies.

The Fourier coefficients, $F(u, \upsilon)$, of the density function are given in the usual way by the 'inverse' transform, which in the present instance is expressed as

$$F(u,v) = \int\!\!\!\int_{-\infty}^{+\infty} f(x,y)\, e^{-2\pi i(ux+vy)}\, dx\, dy \ . \tag{6.09}$$

We need to discover how these Fourier coefficients can be derived from the experimentally measured projections. To do this it seems sensible to rotate the directions of integration in the above equation for the calculation of the Fourier coefficients, so that they coincide with the r,s axes actually used in the experimental work (Fig. 6.02). For a given ϕ we then have

$$[F(u,v)]_{\phi\,\text{const}} = \int_r\!\int_s^{+\infty}_{-\infty} f(x,y)\, e^{-2\pi i wr}\, ds\, dr \tag{6.10}$$

where $\quad w = (u^2 + v^2)^{\frac{1}{2}} \tag{6.11}$

and $\quad \phi = \tan^{-1}(v/u) \ . \tag{6.12}$

w denotes spatial frequency along the r-axis, corresponding to u, v along the x, y axes.

As explained in connection with Eqn (6.04), the integration above, of $f(x,y)$ with respect to s, for a given ϕ, gives the whole 'projection', denoted $p(r,\phi)$, of the section, at that angle ϕ. We can therefore rewrite Eqn (6.10) as

$$[F(u,v)]_{\phi\,\text{const}} = \int_{-\infty}^{+\infty} [p(r,\phi)]_{\phi\,\text{const}}\, e^{-2\pi i wr}\, dr \ . \tag{6.13}$$

$$= [P(w,\phi)]_{\phi\,\text{const}} \ ,$$

where $[P(w,\phi)]_{\phi\,\text{const}}$ recognizes the fact that the integral is the Fourier transform of the projection $[p(r,\phi)]_{\phi\,\text{const}}$. Here, r and w are the conjugates, ϕ is a constant for any one projection, and the capital P denotes the Fourier transform of p (the convention used throughout this book).

We have the important result then, that the Fourier components of a projection $p(r,\phi)$ taken in a given direction ϕ are in fact the Fourier coefficients, $F(u, \upsilon)$, of the actual density function, $f(x, y)$, as a function of r, for that particular direction ϕ. This is an important result because it means that if projections are experimentally obtained in different directions, then a reconstruction of the 2-dimensional density function, $f(x, y)$, can be directly achieved by Fourier synthesis of the Fourier components of all the projections. This overall process of Fourier reconstruction is set out schematically in Fig. 6.05, a representation used by DeRosier and Klug (1968) to describe their work in electron microscopy on improving the imaging of large biological molecules. (Their starting point was quite different from what we have seen here, though the result was the same. They saw that they could invoke a '**projection theorem**' familiar to X-ray crystallographers, and originally due to W. L. Bragg (1929). In this, the Fourier components of a 2-dimensional projection of the 3-dimensional electron density distribution in a crystal are shown to be the central section of the 3-dimensional set of Fourier components of the distribution.)

In the electron microscopy application, photographs (electron 'micrographs') are taken at different angles to the specimen (for example by tilting the latter), and each photograph is analysed into its Fourier components (amplitudes and phases). With careful interpolation to help correct for the angular intervals being too large, the full 3-dimensional transform of the object is obtained. Straightforward Fourier transformation of this gives a recombination of the image (Fig. 5.15).

Independently, Fourier reconstruction in radioastronomy had been discussed by Bracewell in 1956, but calculation of the Fourier integrals was impracticable at that time, causing him to employ an iterative method of the type described in §6.2.3. Since then, developments in computers, and the invention of the **Fast Fourier Transform Algorithm** (Cooley and Tukey, 1965), which exploits the periodic nature of sine and cosine to reduce, drastically, the amount of calculation required in evaluating a Fourier transform, have altered the situation. In X-ray computed tomography the Fourier approach has developed along lines described in the following section.

Although the Fourier transform provides a powerful and natural way of analysing the problem of reconstruction it does not — as we shall see — mean that it is used *explicitly* in all methods. However, as Klug and Crowther pointed out in 1972, 'considering the process in these terms displays its underlying structure in a rather simple way'. They attributed the power of the Fourier approach to the existence of the projection theorem referred to above. Klug and Crowther also formulated the problem in a general way, in the context of **information theory**, and brought out the limitations imposed by having only restricted data available.

(ii) *Filtered back-projection*
The method of 'filtered back-projection' is now widely used in medical computed tomography and is based on the relationship between the Fourier components of the true density distribution, $f(x, y)$, and those of the back-projected distribution, $f'(x, y)$.

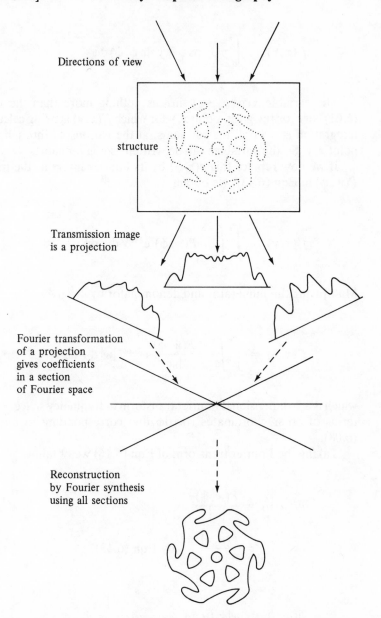

Directions of view

structure

Transmission image
is a projection

Fourier transformation
of a projection
gives coefficients
in a section
of Fourier space

Reconstruction
by Fourier synthesis
using all sections

Fig. 6.05 — Scheme for the general process of reconstruction of a structure from its
transmission images. (After DeRosier and Klug (1968).)

The Fourier components of the true density function are given by Eqn
(6.09), and it remains necessary to obtain the corresponding equation for the
density function obtained by back-projection. To do this we can start with
the expression for the back-projected distribution given by Eqn (6.06) and
rewrite it in integral form as follows:

$$f'(x,y) = \int_0^\pi p(x\cos\phi + y\sin\phi, \phi)\,d\phi \ . \tag{6.14}$$

The variable $x\cos\phi + y\sin\phi$ is nothing more than the value (Eqn (6.01)) of r for the given point x, y for which $f'(x,y)$ is being calculated. The integration is over all the ray sums, at the various ϕ, through the chosen point x, y. So the variables can be written as r, ϕ for short.

If we now replace this $p(r, \phi)$ by its Fourier integral (the transform of $P(w, \phi)$ in Eqn (6.13)) we obtain

$$f'(x,y) = \int_0^\pi\int_{-\infty}^{+\infty} P(w, \phi)\,e^{2\pi i w r}\,dw\,d\phi \ . \tag{6.15}$$

Multiplying the numerator and denominator by $|w|$ gives

$$f'(x,y) = \int_0^\pi\int_{-\infty}^{+\infty} \frac{P(w, \phi)}{|w|}\,e^{2\pi i w r}\,|w|\,dw\,d\phi \tag{6.16}$$

which is a 2-dimensional Fourier transform in frequency space, expressed in terms of polar coordinates ($|w|\,d\phi, dw$) corresponding to du, dv in Eqn (6.08).

Taking the Fourier transform of Eqn (6.16) we obtain

$$F'(u,v) = \frac{P(w, \phi)}{|w|} \tag{6.17}$$

$$= \frac{F(u,v)}{|w|} \qquad \text{by Eqn (6.13)} \tag{6.18}$$

i.e.

$$F(u,v) = |w|\,F'(u,v) \ . \tag{6.19}$$

The Fourier coefficients of the true density function can therefore be obtained by multiplying those of the image obtained by ordinary simple back-projection by $|w|$. By its nature, this procedure for reconstruction can be described as **filtered back-projection**, or **Fourier filtering**.

Although the end result obtained in this way is similar to that obtained by Fourier reconstruction described in (i), this method has been more popular in clinical applications because each projection can be fully processed as it is obtained, instead of waiting until a whole scan (i.e. all the projections) has been completed. For reasons concerned with ease of computation, the algorithms commonly used at present are essentially the convolution equivalent of Eqn (6.19), the method then being referred to as **convolution filtering**, first derived by Bracewell in 1956 in connection with his radioastronomy work, and separately proposed for use in radiography and electron microscopy by Ramachandran and Lakshminarayanan in 1971.

The convolution method was used in the first scanner capable of imaging much larger sections than was the earlier head scanner (§6.2.3). This, the **whole-body scanner**, was developed by Ledley and his colleagues at Georgetown University in the USA and commissioned in 1973 (Ledley *et al.*, 1974). Also to increase the speed and efficiency of those early machines, various changes were made in the scanning geometry, notably from the **parallel-ray geometry** shown in Fig. 6.01(a, b) to the **fan-beam geometry** shown in (c) in that figure. In the newer arrangement, a fan-shaped beam (*ca.* 45°) of divergent X-rays is used with a detector system consisting of some 700 channels, and the combined X-ray source–detector system rotates about the patient to complete the circle. For processing, the data are either re-ordered into a form equivalent to that based on collection from a parallel-ray geometry, or a special convolution algorithm for the fan-beam geometry is used. These machines collect up to *ca* 1000 projections per sec. with up to *ca*

(a)   (b)

Fig. 6.06 — X-ray computed tomography (CT) scans of patient's head. Transverse axial sections (i.e. perpendicular to axis of body).
(a) Showing presence of a cystic tumour.
(b) The same, but following intravenous administration of contrast medium (an iodine-containing compound) which reveals an enhancing solid tumour nodule.
General Electric CT/T 8800. Image 320 × 320 matrix. Fan-beam geometry. Slice thickness 1 cm (Courtesy of Professor I. Isherwood, Department of Diagnostic Radiology, Medical School, University of Manchester, UK).

700 ray integrals/projection, giving an image with a spatial resolution of *ca* 0.5 mm in body sections ('slices') as thin as *ca* 1.5 mm, in about 2 sec. Fig. 6.06 shows scans of a patient's head, revealing the presence of a tumour, and how the contrast in the imaging of this is enhanced by the intravenous administration of an iodine-containing compound.

Dynamic studies of body organs are also possible. A most important example is the investigation of heart-valve action. The normal scanning method of imaging is not possible, however, since the cardiac cycle, typically 800–1000 ms, is short compared with the CT scanning cycle. So it is necessary to gather data in such a way as to allow images to be reconstructed over a series of discrete short intervals in the cardiac cycle. This can be done either by electrocardiogram control of the X-ray tube or by taking the appropriate data, for each interval, from a series of scans performed with simultaneous recording of the electrocardiogram. Such **'gated' computed tomography** makes it possible to put a sequence of images together, for a chosen section through the heart, and to view the sequence as a movie.

6.3   OTHER MODALITIES

6.3.1   Emission computed tomography

Emission computed tomography (ECT) is a technique for obtaining sectional images of the body showing the distribution of artificially made short-lived radioactive isotopes (**radionuclides**). The nuclides are deliberately breathed, ingested or injected into the patient to act as 'tracers'. For instance, the isotope of oxygen ^{15}O (with a half-life of about 2 minutes) emits positrons as it decays and in so doing sends out signals that reveal the pattern of oxygen absorbed in different regions of, for example, the brain. There are numerous examples of how the concentrations of radionuclides in the various organs and structures of the body provide *in vivo* information about normal and abnormal physiological processes. Other useful isotopes for this work include ^{13}N (half-life 10 minutes) and ^{11}C (half-life 20 minutes): all can be used to 'label' a wide variety of pharmaceutical or biological agents involved in metabolic and physiological functions, enabling function rather than structure to be imaged.

ECT methods can be divided into two categories according to how the radionuclides decay:

(1) *Radionuclide decay by direct emission of a γ-ray.*
γ-ray emission computed tomography is alternatively known as **single-photon emission computed tomography (SPECT)**.

(2) *Radionuclide decay by emission of a positron.*
Within *ca* 1 mm the positron interacts with an electron, resulting in their mutual annihilation and the formation of a pair of photons travelling in diametrically opposite directions. This is referred to as **positron emission tomography (PET)**.

Details of the techniques are not relevant here, though it is of interest to note that in PET, pairs of detectors on opposite sides of the body are used so that a signal is recorded only when an annihilation photon is received by each detector. This locates the original site of the positron annihilation to the volume of tissue separating the two detectors. Such '**electronic collimation**' eliminates the need for mechanical collimators which are inherently less accurate. The technique is referred to as **annihilation coincidence detection (ACD)**.

In both categories, image reconstruction is achieved by a convolution (filtered back-projection) algorithm of the same type as that mentioned in connection with X-ray CT (§6.2.4).

6.3.2 Magnetic resonance imaging
I. M. Mackenzie

Within the optical region of the electromagnetic spectrum we are unable — perhaps mercifully — to see inside the human body (though the transparency of some simpler organisms is not unattractive to the human eye). After X-rays and γ-rays, in the high-energy region beyond the ultraviolet, transparency is denied us until we reach beyond the infrared to the low-energy radio region. Here there is an absorption 'window' in human tissue allowing radiation to pass through the body and interact with matter in an interpretable way. This is the domain of **nuclear magnetic resonance (NMR) computed tomography**, now called, by international agreement, **magnetic resonance imaging (MRI)**, to help to avoid it being confused with (X-ray) CT scanning.

Although the use of NMR as a spectroscopic method of chemical analysis goes back more than 30 years, recognition of its potential for actually mapping the spatial distribution of chemical elements in the body is a development of only the past 15 years, largely inspired in the early stages by the work of Lauterbur (1973). Not until the 1980s, however, have commercial magnetic resonance scanners been produced and the method put to clinical use.

The physical phenomenon on which both NMR spectroscopy and MR imaging depend is the magnetic resonance that occurs when atoms with nuclear spin respond to the application of radiofrequency magnetic fields by absorbing or emitting electromagnetic radiation. Table 6.01 lists the principal nuclei of use in the medical field.

The main developments in MR imaging have so far been concerned with the mapping of proton distributions. Hydrogen is the most abundant element in the human body and the technique can distinguish between protons in different environments — water (an indicator of diseased tissue), different types of tissue, etc. Furthermore, as bone is not opaque to radiofrequency radiation the interior of bone cavities are as accessible for imaging as other regions of the body. It is properties such as these that make MR imaging important for its ability to produce information not obtainable by X-ray tomography.

Table 6.01.

Nucleus	Resonance frequency (MHz/tesla)
^{1}H	42.58
^{2}H	6.53
^{13}C	10.71
^{19}F	40.05
^{23}Na	11.26
^{31}P	17.23
^{39}K	1.99

(1 tesla $=$ 10 000 gauss)

When spinning nuclei are subjected to an applied static field, H_0, the couple due to the interaction of their magnetic moments with the field induces a precessional motion, with a frequency (the **Larmor frequency**) characteristic of the nuclear species and proportional to the field strength. H_0 also polarizes the nuclear spin system because the energy of a nucleus is less if its magnetic moment inclines towards the direction of H_0 rather than away from it. Thus the Boltzmann distribution gives rise to a very slight imbalance between the numbers of nuclei which have a positive component of magnetic moment in the direction of H_0 and those with a negative component. The result is a nett magnetic moment M_0 in the direction of H_0 (Fig. 6.07(a)).

The MR imaging technique depends on rotating this nett magnetic moment away from alignment with H_0. Rotation can be achieved by the application of a second, weaker, field H_1 rotating at the nuclear precession frequency in a plane transverse to H_0. The amount of rotation, ϕ in Fig. 6.07(b), depends on the strength of H_1 and the time for which it is applied. And the magnetization M_0, being the sum of the magnetic moments of individual nuclei which are precessing about H_0, itself precesses about H_0. Before the H_1 pulse, M_0 points along H_0 and is stationary: after the pulse, M_0 is inclined to H_0 and its precession about H_0 gives rise to a magnetic moment which is time varying at the natural precession frequency. This time-varying moment will induce an oscillatory flux in a pick-up coil suitably placed near the sample, and so, by Faraday's law, an induced signal voltage. The amplitude of this voltage depends on the initial magnetic moment induced by H_0 and is therefore a measure of the nuclear spin population.

In the recovery after the pulse of H_1 the individual nuclear precessional motions get out of phase with each other; as a result the rotating component of M, namely $M_0 \sin \phi$ (Fig. 6.07(c)), spirals in and dies away to zero, and with it the NMR signal. The randomization of the precessional phases does not affect the component of M_0 along H_0, namely $M_0 \cos \phi$.

The time constant of the decay of $M_0 \sin \phi$ is designated T_2 and is

Fig. 6.07 — (a) The static magnetization M_0 in an applied field H_0.
           (b) An additional rotating field H_1, at the resonance frequency, in-
               duces M_0 to spiral away from H_0 to a chosen value of ϕ.
           (c) The transverse component, $M_0 \sin \phi$, of the magnetization decays to
               zero with a time constant T_2. (During this evolution there will be
               some recovery of the component of M_0 along H_0, i.e. $M_0 \cos \phi$,
               towards its equilibrium value M_0 shown in (a) (the T_1 process). For
               clarity this has not been shown in the diagram.)

variously called the **phase memory time**, the **transverse relaxation time** or the
spin–spin relaxation time. The dephasing is due to the fact that H_0 is not the
only field to which the nuclei in the sample are subject. They are influenced
by the fields caused by the moments of other nuclei (of the same or different
species) or paramagnetic ions in the sample. The random relative motions
and orientations of these moments perturbs the precessional motions of the
nuclei of interest and causes them to get out of phase. It follows that the time
constant T_2 depends on the local environment of the nuclei; indeed some
MR imaging experiments are designed to observe variations in T_2 that occur
in different regions of heterogeneous structures such as the human body.

    As mentioned above, and as can be seen from Fig. 6.07(c), the dephasing
process (the T_2 process) does not, of itself, restore the spin system to its
original equilibrium configuration, i.e. the restoration of the full magnetization M_0 along H_0 as in Fig. 6.07(a). This latter process proceeds with a time
constant T_1 ($> T_2$) called the **spin–lattice relaxation time** ('lattice' because
the original work was done on solids). It involves a loss of energy from the
nuclear spin system (specifically the energy put into the spin system by the
H_1 pulse) and therefore an exchange of energy between the nuclear spins
and their surroundings. The energy ultimately passes into the kinetic energy
of the molecular motion. (The heating effect is negligible because the energy
involved is minute compared with the intrinsic thermal energy of the
sample.)

Procedures have been designed to measure the relaxation times T_1 and T_2 separately. Measurement of T_2 is essentially a measurement of the decay time of the NMR signal, as the magnetization follows the course shown in Fig. 6.07(c). As the NMR signal decays to zero the recovery of the component of M along H_0 towards its ultimate equilibrium value M_0 will be underway (not shown in (c), for clarity) with the longer time constant T_1. If the experiment is repeated within this recovery period the signal will be diminished because the full magnetization along H_0 has not been regained. By varying the time interval between two excitations it is thus possible to build up a picture of the T_1 recovery process.

In MR imaging it is possible not only to observe the distribution of static parameters within a sample, such as proton density, but also to observe the dynamic parameters T_1 and T_2, which provide information about the environment of the nuclei. For protons in biological materials, $T_1 \sim$ 300–1000 ms and $T_2 \sim$ 10–500 ms. A large part of current activity in MR imaging is directed to elucidating the relationship between these dynamic magnetic resonance parameters and the physiological (and pathological) state of tissue.

So far, we have not considered the localization necessary for imaging the actual spatial distribution of nuclei. With the uniform fields implicit in our description up till now, there would be no spatial discrimination in the signals detected: nuclei would respond regardless of their position in the entire volume subjected to the fields. Furthermore, collimation (used, for example, in X-ray tomography) is out of the question since the radiofrequencies employed correspond to wavelengths in the range 10–100 metres. In 1973 Lauterbur had the ingenious idea of superposing linear field gradients on the main field (H_0); 'gradient' meaning that the magnetic field is arranged to vary along a chosen direction, the x-direction say. Then, for a given pulse frequency, only nuclei in a plane at a certain value of x, and perpendicular to x, will have the correct precession frequency to resonate with the pulsed radiation (resonance frequency being proportional to field strength as mentioned earlier).[†] A field gradient applied in different directions in the selected plane (which is, of course, a slice of finite width Δx) then gives a set of projections that can be used to reconstruct an image of the distribution of proton density in the slice (or, with more complicated pulse sequences, of T_1 and T_2).

The fields used in this work range from 200 gauss to 2.3T (1 tesla = 10 000 gauss) according to needs. For a slice, of e.g. 5 mm thickness, across the head (in *any* plane, because the plane whose image is being reconstructed is chosen electronically — a significant advantage over X-ray computed tomography), an image with a resolution of *ca* 2 mm is obtained in a few minutes. Examples are shown in Fig. 6.08. (Up to about sixteen such slices can be imaged in the same overall scan time ('multislice methods')

---

† Lauterbur gave MR imaging the name **zeugmatography** — from the Greek *zeugma*, meaning that which joins together.

(a)   (b)

Fig. 6.08 — Magnetic resonance imaging (MRI) of human head. (a) Coronal section (side-to-side axial plane containing axis body) 128 × 128 matrix. Same patient as in Fig. 6.06. Shows details of tumour only visible in the X-ray CT scan when a contrast medium was used. (b) Mid-line sagittal section (front-to-back axial plane) 256 × 256 matrix. Normal volunteer. Picker International MR system (superconducting magnet). Field 0.26 tesla. Slice thickness 1 cm. (Courtesy of Professor I. Isherwood, Department of Diagnostic Radiology, Medical School, University of Manchester, UK.)

because while one slice is recovering (T_1) measurements can be obtained for another slice.)

MR imaging also offers considerable potential for the imaging of elements other than hydrogen, especially phosphorus (Table 6.01) which is of great interest because of its role in metabolic processes. The molecules of various phosphates, including adenosine triphosphate (ATP), mediate the storage and transfer of energy in the living cell, and measurement of their levels *in situ in vivo* provides the opportunity for unprecedented insight into the condition of the vital organs of the body. Unfortunately, the concentration of ^{31}P is so much less than that of hydrogen that, to obtain adequate sensitivity, much larger volume elements (a few cm^3 of tissue) have to be used, with the consequent loss of spatial volume resolution. The special feature of the technique is that within these, albeit larger, localized volumes, the NMR signals from phosphorus atoms in different chemical environments (e.g. different biological phosphates) can be collected separately. This is because the resonances are modified by the chemical environments of the nuclei. These so-called **chemical shifts** (the basis of conventional **NMR spectroscopy**) are due to shielding effects of the electronic charges in the immediate vicinity of a nucleus, and so depend not only on the atom itself but also on the nature of its bonding to the rest of the molecule.

A furter promising technique is one in which NMR 'tracers' of contrast material are administered, orally or by injection. This is reminiscent of conventional radiography but in the present instance the contrast medium contains paramagnetic ions, such as Mn^{2+}, which have unpaired electrons whose presence modifies relaxation times of NMR-sensitive nuclei, allowing the distribution of the paramagnetic species to be preferentially mapped.

Yet another technique enables the *flow* of material to be measured, for if a nucleus moves up or down the applied field gradient its precessional frequency will alter. When the nucleus reaches a position where the applied field has a particular value its frequency will be the same as, but the phase of its precession will be different from, that of a spin which has been static at that position. These phase differences are at the root of NMR methods used to detect and selectively image, flowing material. Blood flow rates up to $30 \, \text{cm} \, \text{s}^{-1}$ can be measured.

6.3.3 Ultrasonic computed tomography
J. P. Weight

Ultrasound propagates through tissue at velocities typically 5 orders of magnitude less than those of electromagnetic waves. Over the distance ranges of interest in medical imaging it is consequently relatively straightforward to obtain ultrasonic images in a fashion analogous to the pulse-echo technique used to produce pictures by radar. In ultrasonics, such 'plan-position displays' are known as **B-scan images**. They are obtained by scanning a narrow beam in a chosen plane through the body under investigation (as in X-ray CT) but displaying the amplitude of the ensuing ultrasonic reflections versus distance directly as an intensity modulation on (usually) a CRT screen. In this way, beam direction and time-of-flight data provide a 'picture' of the plane under examination. Hence B-scan images can be said to be tomographic in nature. With modern, often electronic, scanning techniques, B-scans can be obtained fast enough to follow the motion of most movements in organs in the body.

The other main form of ultrasonic tomographic image is obtained by reconstruction from transmission projections in fashions entirely analogous to those described earlier in this chapter for X-rays and radionuclide emissions. It is the production of these reconstructed images that we shall refer to here as **ultrasonic computed tomography** (UCT).

There arises the obvious question of why bother with lengthy reconstruction methods when there is already a well established direct imaging technique capable of giving 'real-time displays'? The answer lies in the hope that reconstructed images will eventually contain improved diagnostic information regarding tissue pathology. One of the main problems in obtaining quantitative information from B-scans is that in most instances it has proven difficult to correlate ultrasonic scatter with tissue characteristics. However, a number of studies (see, for instance, Ziskin and Wells, 1980) have shown that variations in ultrasonic *absorption and velocity of propaga-*

tion are often well correlated to tissue properties. Herein lies the interest in UCT, since it is from measurements of just these propagation characteristics that reconstructed images are derived.

As in other forms of CT imaging, data for UCT images are acquired by scanning a plane of the body with a collimated beam so that each point in the scanned plane is interrogated from as many directions as possible. A typical through-transmission system consists of a water tank which contains two coaxially opposed transducers capable of being translated in a linear fashion. The organ to be imaged is immersed and the transducers scanned back and forth whilst the whole tank rotates in order that each point in the image plane is viewed from all directions in the plane. Typical transducers used have centre frequencies of from 1 to 5 MHz with apertures of 1 or 2 cm. Either pencil-like or fan-shaped beam geometries are used with the beam usually being designed to be as narrow as possible in the direction perpendicular to the scan plane. In the pioneering through-transmission work of Greenleaf *et al*. (1974, 1975), typically 10^3 ray-integrals were measured, but more recently, improvements in the speed of reconstruction algorithms have permitted the collection of around 10^4 rays, leading to improved image resolution (Carson and Scherzinger, 1980; Greenleaf, 1983). Usually, pulsed ultrasound is employed, the frequency content, pulse repetition frequency and amplitude of the pulses being similar to those used in B-scan imaging. So far, such levels of ultrasonic insonification have not been shown to cause any harmful effects.

Although reconstruction techniques developed for use in radiography and electron microscopy (especially, Ramachandran and Lakshminarayanan, 1971) are widely used in UCT, some of the assumptions inherent to their valid use are then less than well justified. For instance, such techniques rely on straight line propagation, whereas ultrasonic waves are often severely refracted due to differing velocities of propagation in the rather randomly orientated tissues they pass through. (A somewhat paradoxical situation, since it is just these changes in velocity that can provide some of the most useful tissue characterization information.) Other problems leading to image degradation include the problem of multiple reflections in laminar-like structures and the anisotropic nature of many tissues. Furthermore, it is very difficult to produce clear images in planes which include bones or other hard tissue, due to the very strong reflections ensuing at their surfaces (Carson *et al*., 1977). For these reasons, through-transmission techniques have so far only found application in imaging soft tissues. The requirement of access over ideally, 180° imposes further restrictions on suitable organs for *in vivo* UCT imaging. Up to the present time, the most promising application is for the detection of tumours and other abnormalities in the female breast.

Various ultrasonic propagation properties may be used in reconstructing images, but the two most important are *attenuation* and *velocity of propagation*. Although *ultrasonic absorption* has been shown to be well correlated to certain tissue pathologies, in practice the parameter which is directly measured is the *total ultrasonic attenuation*, which arises by various mechanisms other than by true absorption alone. Thus, scattering, reflection, refraction and beam spreading all contribute to the overall attenuation, and in many situations the loss due to these factors predominates over that due to absorption proper. Although some of these unwanted effects may be compensated for, they still pose major problems to the improvement of attenuation images. Some of these limitations may be circumvented by measuring the time-of-flight of a through-transmitted pulse and reconstructing an image of velocity variations. Currently, most practitioners of UCT use both attenuation and time-of-flight images and further complement their results by displaying a B-scan image of the same tissue plane.

In order to overcome some of the limitations of through-transmission UCT, investigations have been made to explore the feasibility of producing reconstructions from back-scattered ultrasound (Duck and Hill, 1979). This offers the possibility of imaging areas of the body which are not ultrasonically accessible from both sides, as well as the possibility of imaging soft tissues in front of bone. At present, however, further work needs to be carried out in this area in order to overcome the limitations imposed by the need to make a number of simplifying assumptions.

7

Interferometry and radiation sources

7.1 INTRODUCTION

We turn now from the use of radiation for structural studies, image formation and image processing of various types, to the complementary topic of studying the spatial and spectral 'structure' of radiation sources.

In Chapter 1 we have seen in very general terms how the visibility of interference fringes can reveal information about radiation sources. The two types of interferometry pioneered by Michelson are particularly sensitive for such investigations.

Initially, Michelson used his stellar interferometer to measure star diameters, but he foresaw how fringe-visibility measurements might provide information about the brightness distribution of a source and he illustrated this with simple examples.

With his spectral interferometer, Michelson measured the wavelength of spectral lines, but again he foresaw, and demonstrated in simple cases, its greater potential, this time for using fringe visibility to obtain detailed information about the fine-structure of spectra. Later he used it to standardize the metre in terms of the red line of cadmium.

In 1907 Michelson was awarded the Nobel Prize in physics for the development of these optical instruments and for the spectroscopic and metrological investigations which he carried out by means of them.

Today, the two techniques are seen as the archetypes of important methods in modern astronomy and spectroscopy. The principles involved are as important today as they were then. For that reason they are recalled in

§7.2 and §7.3. They also provide a valuable background for an introduction in §7.4 to the modern interpretations based on the theory of partial coherence.

Fourier transform spectroscopy is briefly described in §7.5, and in §7.6 some of the main Fourier aspects of imaging in astronomy, by interferometry, are outlined.

7.2 MICHELSON'S STELLAR INTERFEROMETER

7.2.1 Introduction

To introduce the use of interference effects as an indirect way of using a telescope to measure the angular dimensions of astronomical objects, consider Fig. 7.01(a). This depicts an aperture screen containing two slits

Fig. 7.01 — Measurement of stellar diameters by interferometry.

perpendicular to the figure, and located in front of the lens of a telescope (the same can be done with a reflecting telescope). Wavefronts arrive from all points across a star of **angular diameter** ϕ_0 (the angle it subtends at the earth). Only the extreme wavefronts are indicated in the figure, W_1 originating from one edge of the star disc, W_2 from the opposite edge. In the focal plane of the lens there is a continuous spread of $\cos^2$ intensity fringes (the source being incoherent), ranging from those shown due to W_1 to those due to W_2. The nett result is a fringe pattern, shown in (b), of visibility < 1. Note that the spacing of the fringes stays the same as if the source were a point, viz. $\Delta = f\lambda/D$ (Eqn (1.11)).

In practice the intensity of the fringe pattern is attenuated either side of the axis (cf. the sampling of the single-slit diffraction pattern in §2.4.1). We can ignore this attenuation if the slits are narrow, and particularly if, as in practice, observations are restricted to the central region of the fringe pattern.

Now if D is increased the fringe spacing is decreased and the fringe visibility is reduced, as indicated in (c). When D reaches a value such that the fringe spacing becomes just equal to the spread of the fringes, the visibility becomes zero, as in (d). The fringe spacing is now such that the fringes due to W_2 are displaced by exactly one fringe relative to those due to W_1. In diagram (a) the edge of the star sending wavefronts W_1 is equidistant from B and C, so that if the star were a uniformly luminous rectangle the wavefronts W_2 from the other edge would travel one wavelength further in their journey to C compared with their journey to B. At this special value of $D = D'$ say, we have

$$D'\phi_0 = \lambda \ . \tag{7.01}$$

If D is further increased, the visibility becomes non-zero again, until $D''\phi_0 = 2\lambda$ when it is zero once more, and so on. (These results are of course not peculiar to the special alignment of the W_1 wavefronts with respect to B and C used here for mathematical convenience.)

As a star is effectively a circular luminous disc a correction has to be applied to the above. The method for making this correction is similar to that for deriving the resolving power of a circular aperture (§2.3). The result is that the first disappearance of fringes is not achieved until $D'\phi_0 = 1.22\lambda$, so that the star diameter is given by

$$\phi_0 = \frac{1.22\lambda}{D'} \tag{7.02}$$

or, as a rough guide for yellow light,

$$\phi_0 \approx \frac{0.15}{D'_{(metres)}} \quad \text{seconds of arc} \ . \tag{7.03}$$

The principle of this method, suggested by Fizeau in 1868, was used successfully by Michelson in 1890 to measure the diameters of the moons of the planet Jupiter, obtaining values of around 1 second of arc. For this purpose he used the 12-inch (c. 0.3 m) telescope at Mount Hamilton.

With stars, however, the question arose as to whether atmospheric disturbances and mechanical vibrations of the telescope would prevent measurements from being made because, unlike planets, their diameters were expected to be only a few hundredths of a second of arc. These fears were partially allayed by the successful observation of fringes with the 100-inch (c. 2.5 m) reflector at Mount Wilson. But it was also realized that a telescope large enough to accommodate the separation of apertures necessary for achieving the disappearance of the fringes with such small stellar diameters would need to be 50 ft (c. 15 m) or more in diameter; and with slits at that separation the fringes would become too close to be observable. Furthermore, the result achieved was little different from the resolving power of the same telescope used in the normal way (cf. §2.3).

To overcome these difficulties Michelson devised the stellar interferometer associated with his name. It consisted of a linear array of four mirrors (Fig. 7.02) each about 6 inches (c. 15 cm) in diameter. The outer pair

Fig. 7.02 — The Michelson stellar interferometer.

(M_1, M_2) acted as 'receivers', with separation (what would now be called the **baseline**) adjustable up to 20-ft (c. 6 m). The inner pair (M_3, M_4) were fixed, and they directed the incoming signals through two slits into a telescope to form fringes in the usual way. This scheme allowed the fringe spacing to be maintained at a constant and acceptable value ($\Delta = f\lambda/D$) determined by the separation (D) of the inner mirrors, whilst the path difference required for them to lose visibility was established when the separation L of the outer mirrors reached a value L' such that

$$L'\phi_0 = 1.22\lambda$$

i.e.
$$\phi_0 = \frac{1.22\lambda}{L'} . \tag{7.04}$$

The framework supporting the mirrors was built on to the 100-inch

(c. 2.5 m) Mount Wilson reflecting telescope, now chosen not for its optical resolving power but for its ruggedness as a support for the frame carrying the mirrors. It was essential that they should be rigidly located with respect to each other since very small optical path differences through the system were being detected by the critical adjustment of the separation of the outer mirrors.

With this apparatus Michelson and F. G. Pease measured several giant and supergiant stars whose diameters were smaller than could be measured using the Mount Wilson telescope in the conventional way. For example, they obtained a value of 0.0047 secs of arc for Betelgeuse. (To allow for the darkening towards the 'limb' of a stellar disc they used data for the sun, and on that basis estimated that actual diameters would be about 17 per cent larger than values given by Eqn (7.04).)

In effect the new method gave a 'magnification' of L'/D compared with the first version, and limited only by L in theory. However, the difficulty of ensuring the necessary mechanical stability of a large baseline, together with the effects of turbulence in the atmosphere, prevented any significant further development. In 1930 Pease built a 50-ft beam version but it gave unreliable results and the work was abandoned in 1937.

Michelson realized, however, that the way in which the visibility changes as the separation of the outer mirrors is altered contains information not only about the size but also the brightness distribution of the source. Although the method could not be put to practical use at that time we will look at how he tackled the problem theoretically, because of the importance of the method today in astronomy.

7.2.2 Fringe visibility aspects

Michelson initially considered several examples of sources of different shape but with uniform brightness distribution (Michelson, 1890a). He used a construction essentially like that in Fig. 7.03 (with different symbols) and reasoned along the following lines.

An incoherent source of width W subtends an angle ϕ_0 at the interferometer which has slits perpendicular to the figure and whose separation D is adjustable. (D could equally refer to the separation of the outer mirrors in the later instrument.) Consider light from an elementary strip of width dx in the source at S, distance x from the axis. Let $y = f(x)$ be the length of the strip perpendicular to the figure. The illumination field from this strip is sampled at B and C, and the light diffracted from B and C gives a fringe pattern in the usual way. (The single-aperture 'envelope' of the fringe pattern can be ignored since only fringes very near to the axis need be considered.) To calculate the fringe intensity in direction θ we note that the path difference between the two routes SCE and SB is

$$= (SC + \delta) - SB$$
$$= (SC - SB) + \delta$$
$$= \beta x + \delta \ .$$

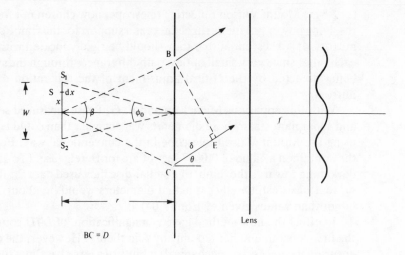

Fig. 7.03

((SC − SB) is evaluated in the same way as in Fig. 2.07, with i in that figure equal here to $\frac{1}{2}\beta$ and assumed small.) The resultant for unit amplitudes is found with the aid of a phasor diagram in the usual way to be

$$R(\theta) = 2\cos\frac{\pi}{\lambda}(\beta x + \delta)$$

which gives the familiar $\cos^2$ type expression for intensity

$$R^2(\theta) = 4\cos^2\frac{\pi}{\lambda}(\beta x + \delta) \qquad (7.05)$$

or

$$R^2(\theta) = 2\left[1 + \cos\frac{2\pi}{\lambda}(\beta x + \delta)\right]. \qquad (7.07)$$

Since the source is incoherent each strip gives a set of separate fringes. And as the strips are of uniform brightness we can multiply the above expression by the area of the strip, $f(x)\,dx$, to represent the weighting of the contribution from that strip to the overall fringe pattern. In the limit when the strip width tends to zero the nett intensity, $I(\theta)$, in direction θ is given by integrating Eqn (7.06)

$$I(\theta) = \int_{-W/2}^{+W/2} f(x)\left[1 + \cos\frac{2\pi}{\lambda}(\beta x + \delta)\right]dx \qquad (7.07)$$

ignoring the factor of 2.

To calculate the fringe visibility according to its usual definition (Eqn (1.05)) it is necessary to obtain the maximum and minimum values of intensity in the resulting pattern. One of the examples Michelson considered was a rectangular source of uniform brightness (Fig. 7.04(a)), whose width

Fig. 7.04.

W was to be determined. The length of the source, $f(x)$, parallel to the slits, was now a constant and could be taken as unity for convenience. The integration in Eqn (7.07) then goes as follows, putting $2\pi/\lambda = k$:

$$I(\theta) = \int_{-w/2}^{+w/2} [1 + \cos k\beta x . \cos k\delta + \sin k\beta x . \sin k\delta] \, dx$$

$$= W + \cos k\delta \int_{-w/2}^{+w/2} \cos k\beta x . \, dx + \sin k\delta \int_{-w/2}^{+w/2} \sin k\beta x . \, dx$$

$$= W + \frac{\cos k\delta}{k\beta} \left[\sin k\beta x \right]_{-w/2}^{+w/2} - \frac{\sin k\delta}{k\beta} \left[\cos k\beta x \right]_{-w/2}^{+w/2}$$

$$= W + \frac{\cos k\delta}{k\beta} (2 \sin k\beta W/2) - 0$$

$$= W + \frac{\cos \dfrac{2\pi\delta}{\lambda}}{\dfrac{2\pi}{\lambda}\beta} \, 2 \sin \frac{\pi\beta W}{\lambda} \, .$$

i.e. $$I(\theta) = W + \frac{\lambda}{\pi\beta} \cos \frac{2\pi\delta}{\lambda} \sin \frac{\pi\beta W}{\lambda} \qquad (7.08)$$

The maximum value of I occurs when $\delta = n\lambda$, where n is zero or an integer, and the minimum occurs when $\delta = (n \pm \frac{1}{2})\lambda$. Allowing for the gross distortion of scale in Fig. 7.03 we can write $r\phi_0 = W$ and $r\beta = D$, giving $\beta W = \phi_0 D$. The following expressions for I_{max} and I_{min} are then obtained

$$I_{max} = W + \frac{\lambda}{\pi\beta}\sin\frac{\pi\phi_0 D}{\lambda} \tag{7.09}$$

$$I_{min} = W - \frac{\lambda}{\pi\beta}\sin\frac{\pi\phi_0 D}{\lambda}. \tag{7.10}$$

Whence the fringe **visibility function**, $V(D)$, is given by

$$V(D) = \frac{I_{max} - I_{min}}{I_{max} + I_{min}} = \left| \frac{\sin\dfrac{\pi\phi_0 D}{\lambda}}{\dfrac{\pi\phi_0 D}{\lambda}} \right|. \tag{7.11}$$

(Denoting visibility as $V(D)$ expresses our interest in V as a function of D for a given λ and a fixed, unknown, value of ϕ_0.)

The sinc function (Fig. 7.04(b)) is familiar in previous chapters as the diffraction pattern, and Fourier transform, of an aperture function having the same 'shape' as the brightness distribution in Fig. 7.04(a). The similarity between the two quite different examples is not a coincidence — but more of that presently (§7.4.1). The visibility curve in (b) drops to zero when $D = D' = \lambda/\phi_0$; and it is zero again when $D = D'' = 2\lambda/\phi_0$, and so on. This is in accordance with the interpretation in the previous section.

Now to take non-uniformity of a brightness distribution into account. If we assume, with Michelson (1891a), that the total intensity from the elemental strip S at x in Fig. 7.03 is given by $B(x)$, this eliminates the added problem of the shape of the source. In effect, it reduces the problem to the determination of a 1-dimensional intensity distribution. Furthermore, if we consider the special case of $B(x)$ as an even function (a brightness distribution that is symmetrical about the axis in the figure) and if the variable x is replaced by ϕ as an angular variable (ranging from $-\phi_0/2$ to $+\phi_0/2$), the following expression for visibility is readily obtained

$$V(D) = \frac{\displaystyle\int B(\phi)\cos(2\pi D\phi/\lambda)\,d\phi}{\displaystyle\int B(\phi)\,d\phi}. \tag{7.12}$$

For the general 1-dimensional case when $B(\phi)$ is not even, though still working in terms of what Michelson called the 'total intensity of a strip of width dx ...', an equation of the following type is obtained

$$[V(D)]^2 = \frac{\left[\int B(\phi) \cos(2\pi D\phi/\lambda)\, d\phi\right]^2 + \left[\int B(\phi) \sin(2\pi D\phi/\lambda)\, d\phi\right]^2}{\left[\int B(\phi)\, d\phi\right]^2}.$$

(7.13)

Michelson concluded, prophetically, that from visibility measurements 'the distribution of luminous intensity of globular masses may be inferred, which would furnish a valuable clue to the distribution of temperature and density in gaseous nebulae'.

An analogous situation arose in Michelson's theoretical study of fringe visibility in relation to spectral distribution (§7.3.2). It was in that connection that Rayleigh (1892) drew attention to the Fourier relationship between fringe visibility and (spectral) intensity distribution. The same applies here, and to conclude this section it will be interesting and informative to pursue the Fourier transform aspect a little further, from a modern viewpoint.

The numerator in Eqn (7.12) is the cosine Fourier transform of $B(\phi)$, with the denominator acting simply as a scale factor. We can readily see that the visibility for each value of the baseline D yields information about one particular Fourier component of the brightness distribution. This is clearly brought out by using the convolution theorem (§4.5). Expressing the observed fringe pattern for a given D as the convolution of $B(\phi)$ with the instrumental response, the convolution theorem tells us that the transform of this convolution is the product of the separate transforms. Now the transform of the instrumental response is a set of $\cos^2$ fringes which have a single spatial frequency determined by the value of D. It follows, therefore, that the transform of the observed fringe pattern for that value of D contains only information about just one harmonic component in the brightness distribution of the source.

The relationship between fringe visibility and coherence is dealt with in §7.4, but first we will look at the complementary topic of fringe visibility as a function of spectral — rather than spatial — distribution in a radiation source, this time using Michelson's spectral interferometer.

7.3 MICHELSON'S SPECTRAL INTERFEROMETER

7.3.1 Introduction

Unlike the stellar interferometer, the spectral interferometer is based on interference by division of amplitude (§1.4). The basic design was given by Michelson in 1881 in connection with an experiment to test whether the

Earth moves relative to an 'aether'. For that purpose he went on to develop the instrument on a larger scale, in conjunction with E. W. Morley (the historical 'Michelson–Morley experiment'). But the basic arrangement was used for the measurement of spectral wavelengths (later to standardize the metre in terms of the red line of cadmium) and the study of spectral fine-structure. It is the spectroscopic applications that remain of significance today, indeed of considerably increased significance.

Figure 7.05(a) shows schematically the arrangement in one of the several

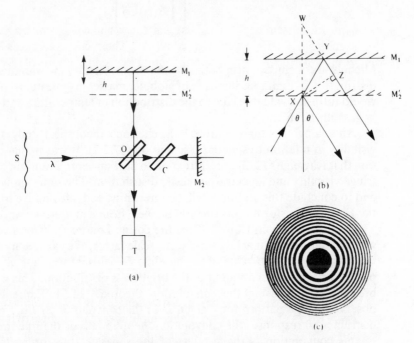

Fig. 7.05 — The Michelson spectral interferometer.
      (a) Basic arrangement (refraction at glass plates O and C not shown).
      (b) Path difference between reflections is $XY + YZ = WZ = 2h \cos \theta$.
      (c) Appearance of the fringes with quasimonochromatic light.

early versions of the interferometer. Light from a source S (usually an extended source) is amplitude-divided at the 'half-silvered' rear surface of a glass plate, O, into two beams, one reflected and one transmitted. The reflected beam travels to the mirror M_1 and is then returned to be partly transmitted through O to the telescope T. Meanwhile, the other beam — the one initially transmitted through the 'beam-splitter' — travels to mirror M_2 and is similarly returned to O, where it is partially reflected to the telescope. Since the beam sent to M_1 passes through plate O three times altogether, compared with the single passage of the beam sent to M_2, a 'compensating plate' of the same thickness and material as O is usually inserted at C. In general, M_1 and M_2 are not equidistant from O and there is an intentional path difference between the two routes (— the compensating plate is only used to equalize the dispersive paths through glass). Reunited, the two beams interfere, giving a resultant that depends on the path difference between them.

M_1 and M_2 are arranged to be mutually perpendicular, and the beam splitter is at 45° to both. The image of M_2 in O, as seen with the telescope, is at M_2', parallel to (or coincident with) M_1. The interference pattern observed with the telescope is therefore like that observed with the parallel plate in Fig. 1.09, though in the present example it is equivalent to being formed by reflection at an 'air plate'. With an extended source, of wavelength λ, rays can enter the system over a wide angular range and concentric bright rings are seen (Fig. 7.05(c)) (cf. Fig. 1.09(b)). Their circumferences are in directions θ for which there is reinforcement between the pairs of recombining wavetrains. That condition is given by

$$2h \cos \theta = m\lambda \qquad (7.14)$$

where m is an integer or zero, and h is the mirror 'separation' (Fig. 7.05(b)). This assumes that the two interfering beams undergo identical phase changes at the beamsplitter. If that is not so, a constant has to be added to the phase difference associated with the path difference, and all fringes are displaced accordingly.

One mirror, M_1 in the figure, can be translated parallel to itself in the direction shown. Altering h causes the ring pattern to expand or contract; if h increases, the rings spread out from their centre as though generated there, and if h is decreased they shrink in to the centre.

An expression for the radial intensity distribution out from the centre of the pattern for a given value of h and wavelength λ is readily derived in the usual way with a phasor diagram. If the amplitudes of the light arriving at the telescope from the two routes are arranged to be equal, A say, then the resultant intensity, I_1, in direction θ in the ring system is found to be given by

$$I_1 = 4A^2 \cos^2 (\psi/2) \qquad (7.15)$$

where the phase difference

$$\psi = \frac{2\pi}{\lambda} 2h \cos \theta \ . \qquad (7.16)$$

Therefore we have

$$I_1 = 4A^2 \cos^2 \left(\frac{\pi}{\lambda} 2h \cos \theta \right) \ . \qquad (7.17)$$

The fringes for ideally monochromatic light are therefore of the $\cos^2$ type, as illustrated in Fig. 7.06(a). Furthermore, from what we have noted

Fig. 7.06 — (a) Cos^2 fringes. (b) Michelson's results for Hg(λ 5461 Å). (Adapted from Michelson, 1927.)

above about the effect on the ring pattern of altering h, it follows that if h is progressively increased or decreased, a detector at any point in the pattern (it could be on the axis, i.e. $\theta = 0$) would register a sinusoidal variation of intensity. If the light were indeed strictly monochromatic the wavetrains would be of infinite length (§4.6), and the sinusoidal pattern of fringe visibility would be quite unaffected by any amount of path difference introduced between the interfering light beams. Conversely, if such a pattern were experimentally observed one could deduce that the light was of such monochromaticity. If, in contrast, with another light source the visibility dropped to zero almost as soon as a path difference was introduced, it could be concluded that the light from the source had a broad continuous spectrum since the wavetrains must be very short (§4.6). It is the quantitative application of that philosophy to the analysis of optical spectra that is the key to the use of the interferometry method.

Consider for the moment a further hypothetical example. Suppose that the light being examined consists of two closely separated, ideally monochromatic wavelengths, λ_1 and λ_2. The changing pattern of intensity registered by our detector is now more complicated than in the above example of a single monochromatic wavelength. For the chosen θ-position of the detector there will be values of h for which rings of the two systems almost, if not totally and exactly, coincide: the detector will register a stronger signal. This happens when $h = h_1$ say, such that

$$2h_1 \cos \theta = p\lambda_1 \approx q\lambda_2$$

where p and q are integers. (In practice the two ring-systems can, for such a value of h, be almost in coincidence over quite a wide angular range if $\lambda_1 \sim \lambda_2$ is small.)

Increasing (or decreasing) h causes the two sets of rings to separate again, however slightly, and the detector records a succession of lower-intensity maxima and non-zero minima as they pass it. The details of the changing signal will depend on the difference between the two wavelengths, their relative intensities and, in real examples, on the line shapes and fine-structure. Since the two ring-systems move out (or in) from the centre of the pattern at different rates (cf. Eqn (7.14)) a new value of h is reached, $h = h_2$ say, for which 'coincidence' again occurs and the signal at the detector rises again; one set of rings has overtaken the other by one whole fringe interval. This condition can be expressed as

$$2h_2 \cos \theta = (p + k)\lambda_1 \approx (q + k + 1)\lambda_2$$

where k is some number.

From the above equations we find, by subtraction, that

$$2(h_2 - h_1) \cos \theta = k\lambda_1 = (k + 1)\lambda_2 \ .$$

This result can be used in various ways. For example, suppose that λ_1 is known and the observations are made on or near the axis, i.e. $\cos \theta \approx 1$. Put $2(h_2 - h_1) = H$. Then

$$\lambda_2 = H\lambda_1/(H + \lambda_1) \ .$$

Measurement of H therefore enables λ_2 to be calculated.

The considerable accuracy of this method for determining λ_2 is illustrated by a rough calculation that shows that the difference in wavelength between the two 'lines' in the sodium doublet (λ 589.0 nm and 589.6 nm) necessitates a mirror displacement of about 0.3 mm between successive coincidences. (We will not digress into the question of the resolving power as such, since our interest is really in the Fourier method described in §7.5.)

This way of using the interferometer is analogous to the earlier observation by Fizeau (1862) that by the 500th order in a Newton's rings experiment using a sodium source, the rings had almost disappeared (i.e. zero visibility), but that they regained their clarity at the 1000th order. He deduced that the sodium light was a doublet, with the 1000th order ring of the longer wavelength coinciding with the 1001th order ring of the shorter, the difference in wavelength between the two therefore being about 1/1000 of their mean.

Michelson realized, however, that much information is lost in this method of analysis. He made visual estimates (transferred to a quantitative scale by a separate, ingenious calibration experiment) of fringe visibility as a function of mirror displacement. He realized that this 'visibility curve' contained very detailed information about the spectrum of the light source.

As early as 1887 Michelson had shown by careful observation and reasoning that 'the red hydrogen-line is a very close doublet; and the same is true for the green thallium-line'.

His mathematical analysis of these matters, together with an important contribution by Lord Rayleigh shortly afterwards in a published letter (1892), concerning the Fourier transform relationship between visibility and spectral distribution, are outlined in the following section. (The reader can, however, proceed straight to §7.5.)

7.3.2 Fringe visibility and spectral distribution

Michelson's research (1891b) under this heading is (dare one say) illuminating and important in its modern relevance. The significant opening statement on the matter was: 'The general formula for the visibility of fringes due to interference of two streams of light whose difference of path is variable, from a source which is not homogeneous, is the same as that for a source of finite area whose "parallax" is variable'. (For 'homogeneous' we would say monochromatic.)

In other words, the suggestion was that the relationship between fringe visibility and spectral distribution is the same as that between fringe visibility and the spatial distribution of a source.

He started with the following expression for the intensity in direction θ in the fringe pattern obtained with the spectral interferometer, for light of wavelength λ:

$$I_1 = 4\cos^2\left(\frac{\pi l}{\lambda}\right)$$

where l = path difference ($2h\cos\theta$ in Eqn (7.17)).

In spectroscopy the convention is to use wavenumber ($\sigma = 1/\lambda$) rather than wavelength. The above equation is then written as

$$I_1 = 4\cos^2\pi l\sigma . \tag{7.18}$$

If the light is polychromatic the fringe intensities in direction θ due to different wavenumbers are additive as there is no coherence between them. If a spectral intensity distribution $B(\sigma)$ extends from σ_1 to σ_2, the total intensity in direction θ is given by

$$I = \int_{\sigma_1}^{\sigma_2} I_1 \, d\sigma = 4\int_{\sigma_1}^{\sigma_2} B(\sigma)\cos^2\pi l\sigma \, d\sigma . \tag{7.19}$$

For σ substitute $\bar{\sigma} + \sigma'$ where $\bar{\sigma}$ is the mean wavenumber.

Replacing $B(\sigma)$ by $f(\sigma')$ then gives (omitting the numerical factor)

$$I = \int_{-a/2}^{+a/2} f(\sigma')\cos^2\pi l(\bar{\sigma} + \sigma') \, d\sigma' \tag{7.20}$$

where the spectral range from σ_1 to σ_2 is now expressed as being from $\sigma' = -a/2$ to $+a/2$ about the mean.

Expanding the $\cos^2$ term and following Michelson by putting

$$\int f(\sigma')\,d\sigma' = P\,, \qquad \int f(\sigma')\cos 2\pi l\sigma'\,d\sigma' = C$$

$$(7.21)$$

$$2\pi l\bar\sigma = \theta\,, \qquad \int f(\sigma')\sin 2\pi l\sigma'\,d\sigma' = S$$

gives

$$I = P + C\cos\theta - S\sin\theta\,.$$

(Some changes in symbols have been made.)

If a is sufficiently small, the variations of C and S with θ may be neglected, and the maxima and minima of intensity occur when

$$\tan\theta = -S/C \qquad\qquad (7.22)$$

which gives $\qquad\qquad I = P \pm \sqrt{C^2 + S^2}$

and therefore $\qquad\qquad V^2 = (C^2 + S^2)/P^2\,. \qquad\qquad (7.23)$

Michelson noted that this is of exactly the same form as the expression for visibility associated with a spatially extended source. Equation (7.13) referred to a brightness distribution $B(\phi)$, whereas here we have a spectral intensity distribution $f(\sigma')$.

It was in connection with Eqns (7.21) that Rayleigh (1892) pointed to the Fourier relationship involved. At the same time he drew attention to the difficulty of using it to deduce a spectral distribution from visibility measurements because of the lack of information about the phases of the harmonics comprising the distribution: '... the visibility-curve by itself gives, not both C and S, but only $C^2 + S^2$; so that we must conclude that in general an indefinite variety of structures is consistent with a visibility-curve ...'.

Despite this limitation, Michelson found it possible to deduce the structures of some simple spectral lines, and these were broadly confirmed by subsequent work.

The expressions for C and S in Eqns (7.21) are the coefficients in a Fourier representation of $f(\sigma')$ which Michelson wrote as

$$f(\sigma') = \int C \cos 2\pi l \sigma' \, dl + \int S \sin 2\pi l \sigma' \, dl \ .$$

Noting from (7.22) and (7.23) that $C = PV \cos \theta$ and $S = PV \sin \theta$ gave

$$f(\sigma') = \int V \cos (2\pi l \sigma' + \theta) \, dl \ . \qquad (7.24)$$

If $f(\sigma')$ is symmetrical (remember that σ' is wavenumber relative to the mean value $\bar{\sigma}$), then $S = 0$ and

$$f(\sigma') = \int V \cos 2\pi l \sigma' \, dl$$

which allowed $f(\sigma')$ to be found by integration using the experimentally obtained visibility curve, V (or $V(D)$ as we denoted it earlier). In the case of an unsymmetrical $f(\sigma')$ it was necessary to know V and θ, the latter (which is also a function of l) being called the 'phase curve' by Michelson. In a number of examples he made estimates of θ by comparison with an 'approximately homogeneous symmetrical source of approximately the same wavelength'. (It is interesting to reflect that for this work Michelson developed his own mechanical sine and cosine analogue computer, which he called a 'harmonic analyser'.) Figure 7.06(b) shows an example of his investigations. It is the visibility curve for mercury (λ 5461 Å), and the deduced spectral distribution in which the two main conponents were estimated to be 0.1 Å apart (1 Å = 0.1 nm).

The development of the full potential of the method, however, had to await the advent of the modern digital computer. In its application for wavelength measurement the two-beam interferometer was superseded by the multiple-beam method used in the Fabry–Perot interferometer. Then in the 1950s it started to make its return, to become the basis of modern Fourier transform spectroscopy (§7.5).

7.4 PARTIAL COHERENCE, CORRELATION, AND VISIBILITY
(not essential for §7.5)

The following treatment of partial coherence is essentially a simplified account of that given by Born and Wolf in their *Principles of Optics*, and largely based on the work of Zernike (1938) and Hopkins (1951). Unlike earlier studies, it is more closely related to experiment.

In Fig. 7.07 consider the illumination field produced in plane C by a source, W, that is both extended and polychromatic. Two points, C1 and C2, are chosen in the field, and by positioning a screen with a pinhole at each the visibility of interference fringes produced by the light from those particular

Fig. 7.07

places in the field can be experimentally measured. We seek the relationship that allows the temporal and spatial coherence of the illumination to be calculated from such measurements. After all, the fringe visibility is but a contrived physical manifestation of the coherence and we hope to relate it analytically to the more fundamental property.

Consider the interference at some general (off-axis) position of P. There is a time delay, τ say, between the journey times C1–P and C2–P. Without losing generality this difference can be assigned (solely to simplify the equations we write) to one path. If the complex amplitudes of the illumination at C1 and C2 are $A_1(t)$ and $A_2(t)$ respectively, the intensity at P can be expressed as

$$I_P = \ <[A_1(t+\tau)+A_2(t)][A_1^*(t+\tau)+A_2^*(t)]> \tag{7.25}$$

where * denotes the complex conjugate. (To simplify this presentation complex amplitude transmission factors ('propagators') for the paths through the apertures, that take into account the size and distance of the apertures from P, have been omitted.)

The $<\ >$ brackets in this equation are used to denote that measurement of intensity is (normally) a time-averaging process†. The averaging is assumed to be over a time-span greater than the coherence time of the light.

Multiplying out gives

$$I_P = \ <A_1(t+\tau)A_1^*(t+\tau)> + <A_2(t)A_2^*(t)>$$
$$+ <A_1(t+\tau)A_2^*(t)> + <A_1^*(t+\tau)A_2(t)> \ . \tag{7.26}$$

The first two terms are the intensities at C1 and C2 respectively, which

---

† The **time-average** of a function $f(t)$ taken over an interval T is given by

$$<f(t)> \ = \frac{1}{T}\int_t^{t+T} f(t')\,dt'$$

where t' is a dummy variable.

can be denoted by I_P^{C1} and I_P^{C2}. (In the first term, τ simply shifts the time origin of the averaging and this normally has no effect on the value of the average: the system is assumed to be 'stationary', i.e. unchanging on the time-scale involved here.) The sum of the third and fourth terms is the sum of a complex number and its own complex conjugate, and is therefore just twice its real part. The equation can therefore be written as

$$I_P = I_P^{C1} + I_P^{C2} + 2\mathscr{R}(\Gamma_{12}(\tau)) \qquad (7.27)$$

where $\mathscr{R}(\Gamma_{12}(\tau))$ is the real part of

$$\Gamma_{12}(\tau) = <A_1(t+\tau)A_2^*(t)> . \qquad (7.28)$$

The time-averaging in this definition of Γ_{12} can be expressed as an integration, and it is then obvious by reference to §4.7 that $\Gamma_{12}(\tau)$ can be described as the **complex cross-correlation function** between the illumination field at C1 and C2, in which the vibrations at C1 are 'considered' at time τ later than at C2. In the present context of physical optics $\Gamma_{12}(\tau)$ is also commonly referred to as the **complex mutual coherence function** of the illumination field at those points. $\mathscr{R}[\Gamma_{12}(\tau)]$ is then called the **mutual coherence**, and by comparison with our analysis of double-aperture diffraction in §1.1 its role in Eqn (7.27) clearly corresponds to what we recognized as the 'interference term' in Eqn (1.01).

When C2 coincides with C1, Eqn (7.28) becomes

$$\Gamma_{11}(\tau) = <A_1(t+\tau)A_1^*(t)> \qquad (7.29)$$

which is the **complex self-coherence** (or **autocorrelation**) **function** of the field at C1. For $\tau = 0$, Γ_{11} reduces to just the intensity due to C1 alone, and in this simplified treatment we can write

$$I_P^{C1} = \Gamma_{11}(0)$$

$$\qquad (7.30)$$

$$I_P^{C2} = \Gamma_{22}(0) .$$

Before proceeding, it is useful to normalize expressions such as Eqn (7.28), to enable coherence to be specified independently of the moduli of the amplitudes — they are really irrelevant to such a statement. The normalized complex coherence function is defined by the **complex *degree* of mutual coherence** (or **cross-correlation**), $\gamma_{12}(\tau)$, as follows

$$\gamma_{12}(\tau) = \frac{<A_1(t+\tau)A_2^*(t)>}{[<|A_1|^2> <|A_2|^2>]^{\frac{1}{2}}} = \frac{\Gamma_{12}(\tau)}{\sqrt{\Gamma_{11}(0)}\sqrt{\Gamma_{22}(0)}} . \qquad (7.31)$$

It is known as the **phase coherence factor**.

Using the Schwarz inequality in mathematics it can be shown that

$$0 \leqslant |\gamma_{12}(\tau)| \leqslant 1 .$$

Eqn (7.27) then becomes

$$I_P = I_P^{C1} + I_P^{C2} + 2\sqrt{I_P^{C1}}\sqrt{I_P^{C2}}\mathcal{R}(\gamma_{12}(\tau)) \tag{7.32}$$

where $\mathcal{R}(\gamma_{12}(\tau))$ denotes the real part of $\gamma_{12}(\tau)$.

This equation is variously known as the *general interference law for stationary optical fields*, or as the *general interference law for partially coherent light*. An important feature is that it permits $\mathcal{R}(\gamma_{12}(\tau))$ to be calculated from experimentally measurable quantities. For any chosen pair of points in an illumination field a screen with suitably located pinholes is used just as in Fig. 7.07. The intensity, I_P, is measured at various positions of P to give an appropriate range of τ; and the intensity at P from each pinhole alone, I_P^{C1} and I_P^{C2}, is also measured. The three quantities give $\mathcal{R}(\gamma_{12}(\tau))$ for each value of τ:

$$\mathcal{R}(\gamma_{12}(\tau)) = \frac{I_P - I_P^{C1} - I_P^{C2}}{2\sqrt{I_P^{C1}}\sqrt{I_P^{C2}}} . \tag{7.33}$$

The precise meaning of $\mathcal{R}(\gamma_{12})$ is made clearer by writing $\gamma_{12}(\tau)$ itself as

$$\gamma_{12}(\tau) = |\gamma_{12}(\tau)|\, e^{i\phi_{12}(\tau)} \tag{7.34}$$

where $\phi_{12}(\tau)$ refers to the generalized phase angle between the fields at P. We can then write

$$\phi_{12}(\tau) = \alpha_{12}(\tau) + \delta \tag{7.35}$$

where δ is the phase angle associated solely with the internal path difference between C1 $\rightarrow$ P and C2 $\rightarrow$ P.

The real part of $\gamma_{12}(\tau)$ in Eqn (7.34) can now be expressed as follows

$$\mathcal{R}(\gamma_{12}(\tau)) = |\gamma_{12}(\tau)| \cos[\alpha_{12}(\tau) + \delta] \tag{7.36}$$

and Eqn (7.32) becomes

$$I_P = I_P^{C1} + I_P^{C2} + 2\sqrt{I_P^{C1}}\sqrt{I_P^{C2}}\,|\gamma_{12}(\tau)|\cos[\alpha_{12}(\tau) + \delta] . \tag{7.37}$$

The similarity to Eqn 1.01 is now even closer, and by comparing the two it is apparent that if $|\gamma_{12}(\tau)|$ has its upper extreme value, unity, then the intensity at P is the same as would be obtained with ideally monochromatic light and with a phase difference between the wavemotions at C1 and C2 equal to $\alpha_{12}(\tau)$. Under those circumstances the illumination field at C1 and C2 would be coherent. If $|\gamma_{12}(\tau)|$ has its other extreme value, zero, the final term in

Eqn (7.37) is absent; there are no interference effects, because the illumination field at C1 and C2 is incoherent, and the observed intensity is simply the sum of the independent intensities due to C1 and C2. Most importantly, intermediate conditions, i.e. **partial coherence**, can be quantified by $|\gamma_{12}(\tau)|$, the **degree of mutual coherence**.

This treatment leads to a convenient concept of partial coherence as being equivalent to a mixture of coherent and incoherent light with intensities in the ratio

$$\frac{I_{\text{coh}}}{I_{\text{incoh}}} = \frac{|\gamma_{12}(\tau)|}{1 - |\gamma_{12}(\tau)|} . \tag{7.38}$$

However, the basically statistical nature of the physical phenomena that are being modelled in the whole of this treatment must be firmly remembered.

Returning to Eqn (7.37), we have not as yet seen how to deduce the modulus and argument of γ_{12} from experimental measurements; we have one equation and two unknowns. Let us reconsider our position. At the outset, and in order to obtain a general picture, the source was postulated as being both spatially and spectrally extended. All our considerations so far have respected that, and as a result the various equations referring to γ_{12} have been quite general with regard to the coherence of the illumination. Now Fig. 7.07 refers to a comparison between different points in the sampling plane of C1 and C2. This is clearly an arrangement that is especially sensitive to spatial (transverse) coherence. To relate γ_{12} to observables it is sensible, then, to consider the case where temporal coherence is not a complicating factor and we do this in §7.4.1. It is $\Gamma_{11}(\tau)$ that is appropriate for dealing specifically with temporal coherence since it is concerned with the extent to which phase relationships are maintained along individual wavetrains and we consider that in §7.4.2.

7.4.1 Spatial coherence

Assume that the extended source in Fig. 7.07 is quasimonochromatic to the extent that the coherence time is much greater than the maximum value of τ associated with P for the range of positions of C1 and C2 over which we wish to assess the spatial coherence in the illumination field. Under those circumstances Eqn (7.37) gives maximum and minimum values of I_P when $\cos[\alpha_{12}(\tau) + \delta] = \pm 1$, since with the above assumption the background intensity level (and other factors dealt with in a rigorous analysis) will change only slowly through the fringe pattern. Furthermore, if τ itself is quite small it can be shown that the degree of mutual coherence, $|\gamma_{12}(\tau)|$, is inappreciably different from $|\gamma_{12}(0)|$, i.e. V and γ are largely independent of τ. Using all this gives the following expression for visibility

$$V_{12} = 2|\gamma_{12}| \frac{\sqrt{I_P^{C1}} \sqrt{I_P^{C2}}}{I_P^{C1} + I_P^{C2}} . \tag{7.39}$$

This is a relationship we have sought, and if the individual intensities associated with C1 and C2 are equal — as is often the case — then the *fringe visibility is equal to the degree of mutual coherence (cross-correlation)*, i.e.

$$V_{12} = |\gamma_{12}| . \tag{7.40}$$

Of Eqn (7.36) we need to note that the phase component α_{12} of the mutual coherence (i.e. any phase difference between the fields at C1 and C2) is revealed by the amount of off-axis shift in the position of the central fringe maximum, since the on-axis position corresponds to $\delta = 0$.

Experimental measurement of fringe visibility and fringe position therefore provides direct information about both the modulus and argument of the complex degree of spatial coherence in the illumination field given by an extended source.

With this direct relationship between visibility and correlation in mind we return to the similarlity noted in §7.2.2 between two Fourier pairs, viz. the visibility–brightness distribution pair in Fig. 7.04 and the diffraction pattern–aperture function pair so familiar from previous chapters. As mentioned at the time, the similarity is not fortuitous or peculiar to that particular example. It can be shown that what one might call the 'pattern' of the complex degree of coherence (cross-correlation) in a plane illuminated by an extended source is the same as the complex-amplitude diffraction pattern given by an aperture of the same size and shape as that source. This is expressed formally in the **van Cittert–Zernike theorem**, to be found in the more advanced textbooks. Apart from its theoretical significance the theorem is particularly important from a computational point of view because the coherence calculation is normally more difficult than the calculation of the corresponding diffraction pattern.

Most importantly, a brightness distribution can be computed by Fourier transformation of the cross-correlation function derived from fringe visibility and phase data. And from our study of the spectral interferometer it is to be expected that a similar relationship exists between autocorrelation and spectral distribution. That is dealt with in the following section.

7.4.2 Temporal coherence

Here we show that the Fourier transform of the **complex self-coherence** (or **autocorrelation) function**, $\Gamma_{11}(\tau)$, is the spectral intensity distribution (the 'power spectrum'). We start by rewriting the time-averaging definition of $\Gamma_{11}(\tau)$ in Eqn (7.29) as an integration as follows

$$\Gamma_{11} = \int A(t + \tau)A^*(t)\, dt \tag{7.41}$$

where the earlier subscripts to A and A^* are implied. Then, by definition, the transform of $\Gamma_{11}(\tau)$ is

$$T[\Gamma_{11}(\tau)] = \int \Gamma_{11}(\tau) \, e^{2\pi i \nu \tau} \, d\tau \qquad (7.42)$$

where ν and τ are the frequency–time conjugate pair.

Substituting for Γ_{11} from Eqn (7.41) gives

$$T[\Gamma_{11}(\tau)] = \int_\tau \int_t A(t+\tau) \, A^*(t) \, e^{2\pi i \nu \tau} \, dt \, d\tau$$

$$= \int_t A^*(t) \, e^{2\pi i \nu t} \, dt \int_\tau A(t+\tau) \, e^{-2\pi i \nu (t+\tau)} \, d\tau \qquad (7.43)$$

$$= T[A^*(t)] \, T[A(t)] \, . \qquad (7.44)$$

(The second integral in Eqn (7.43) is with respect to τ and is independent of t since the system is assumed to be stationary — cf. the discussion of Eqn (7.26).) Putting $F^*(\nu) = T[A^*(t)]$ and $F(\nu) = T[A(t)]$ we obtain

$$T[\Gamma_{11}(\tau)] = F^*(\nu) \, F(\nu)$$

$$= |F(\nu)|^2 \, . \qquad (7.45)$$

This equating of the Fourier transform of the autocorrelation function with the intensity (power) spectrum is an expression of the **Wiener–Khintchine theorem** (cf. §4.7.1).

The relationship parallels the result of the Michelson–Rayleigh treatment in §7.3.2. $|\Gamma_{11}(\tau)|$, or its normalized equivalent, the degree of **self-coherence**, $|\gamma_{11}(\tau)|$, can be related to visibility just as for $|\gamma_{12}|$ in Eqn (7.39); and the spectral intensity distribution, $|F(\nu)|^2$, was denoted in §7.3.2 by $f(\sigma')$. Visibility curves obtained with the two-beam spectral interferometer may be interpreted as representing $|\gamma_{11}(\tau)|$, a function of the path difference (corresponding to time τ) introduced in the comparison of a wavetrain with itself.

$$\bullet \quad \bullet \quad \bullet \quad \bullet \quad \bullet$$

The elegance and 'symmetry' of the two pairs of Fourier transform relationships in interferometry are strikingly evident now. The visibility curve obtained in the spectroscopic application is in the time domain, i.e. it is a function of the time delay introduced between the two optical paths in the spectral interferometer in which a wavetrain is compared with itself (autocorrelation): the transform is the intensity ('power') spectrum of the source. With the stellar ('spatial') interferometer the visibility curve is a function of distance between two points in an illumination field that are compared (cross-correlation): its transform is the brightness spatial (angular) distribution of the source.

7.5 FOURIER TRANSFORM SPECTROSCOPY

In §7.3.1 we saw that with the Michelson spectral interferometer every wavetrain of light is amplitude-divided into two wavetrains that are then brought together after a path difference has been introduced between them. We saw that the way in which the visibility of the interference fringes changes as the path difference is altered (by translating mirror M_1 in Fig. 7.05) contains information about the spectral composition of the light. The routine practical application, however, had to await the arrival of the modern computer. Now, particularly because of the progress made in recent times in the development of small, dedicated computers and microprocessors as integral components of laboratory equipment, the advantages of 'Fourier spectroscopy' are being widely realized.

In this section we consider the Fourier aspect in more detail, in the context of some present-day techniques and applications.

The arrangement used in some types of modern Fourier transform spectrometer is shown in Fig. 7.08. It differs from that in Fig. 7.05 in one

Fig. 7.08 — Fourier transform spectrometry.

major respect; light from the source is collimated by a mirror C before being amplitude divided by the beamsplitter B. This is the Twyman–Green version of the Michelson. The collimation has the effect of making the whole cross-section of the illumination field through the instrument correspond to the axial ($\theta = 0$) direction in Fig. 7.05. Circular fringes are therefore absent and the whole field is of uniform intensity. The changes in this intensity that occur as M_1 is translated are measured with the mirror and detector arrangement shown in the figure. Thus, in the case of the hypothetically monochromatic light that we considered, the detector would again record a sinusoidally varying intensity of illumination. If the wave-number is σ_1 and the combining beams are of equal amplitude, A_1, say, then the intensity as a function of l is given by

$$I_{\sigma_1}(l) = 4A_1^2 \cos^2(\pi\sigma_1 l) .$$

corresponding to Eqn (7.17) with the path difference $l = 2h$, since $\theta = 0$. This can be written as

$$I_{\sigma_1}(l) = 2A_1^2(1 + \cos 2\pi\sigma_1 l) . \tag{7.46}$$

The intensity measured by the detector for light of spectral intensity distribution $B(\sigma)$ is therefore given by

$$I(l) = \int_0^\infty B(\sigma)(1 + \cos 2\pi\sigma l)\, d\sigma \tag{7.47}$$

omitting the numerical factor.

This integral can be written as

$$I(l) = \int_0^\infty B(\sigma)\, d\sigma + \int_0^\infty B(\sigma) \cos(2\pi\sigma l)\, d\sigma . \tag{7.48}$$

When the two mirrors are exactly equidistant, optically, from the beamsplitter, $l = 0$ and $I(l) = I(0) = 2\int B(\sigma)\, d\sigma$. Therefore we have

$$I(l) - \tfrac{1}{2}I(0) = \int_0^\infty B(\sigma) \cos(2\pi\sigma l)\, d\sigma . \tag{7.49}$$

This is a cosine Fourier transform relationship and we can write

$$B(\sigma) = \int_0^\infty [I(l) - \tfrac{1}{2}I(0)] \cos(2\pi\sigma l)\, dl . \tag{7.50}$$

This equation, derived more rigourously in the specialist literature, is often described as the basic equation of Fourier transform spectroscopy. It enables the intensity spectrum, $B(\sigma)$, to be calculated for each chosen value of σ by performing the integration on the right.

Let us see what is involved. $I(l)$ is the intensity measured as a function of l. Plotted graphically it is called the 'interferogram', and Fig. 7.09(a) shows an example. (We shall discuss negative values of l presently.) Now in any

Fig. 7.09 — Interferogram (a) and calculated spectrum (b). (Adapted from Bell (1972).)

interferogram, for $l = 0$ there is no path difference between combining wavefronts; amplitudes are added, and $I = \mathrm{I}(0) = \sum (2A)^2 = 4 \sum A^2$. At the other extreme, when l is so large that interference can no longer take place (i.e. l is greater than the coherence length of the light), then it is separate intensities that are added, giving $2 \sum A^2 = \frac{1}{2}I(0)$. Inspection of Fig. 7.09(a) in the light of these identifications shows that $[I(l) - \frac{1}{2}I(0)]/\frac{1}{2}I(0)$ would be the 'visibility curve' (visibility as defined previously). We can therefore say that Eqn (7.50) relates $B(\sigma)$ to the Fourier transform of the visibility curve derived from the interferogram. Figure 7.09(b) shows the result for (a).

Note that in Eqn (7.50) the calculation of $B(\sigma)$ for each chosen wavenumber, σ_1 say, amounts to multiplying the whole of the curve $[I(l) - \frac{1}{2}I(0)]$

by $\cos(2\pi\sigma_1 l)$, the area under the product curve giving the required value of $B(\sigma_1)$. This is how the curve in Fig. 7.09(b) is obtained — though with a computer. It is reminiscent of the determination of Fourier coefficients in Chapter 3 (see Fig. 3.02) except that we are now dealing with non-repeating 'patterns' and therefore an integral rather than a series is involved. Also, we are working in the frequency–time domain, represented by σ and l respectively (l corresponds to a time delay).

Experimental measurements for negative values of l are as easy to make as for l positive though the interferogram should be symmetrical. If, for any reason, it was not symmetrical, the cosine Fourier transform would not be applicable because it implicitly assumes that $I(l)$ is an even function. For reasons that are outside the scope of this book some instruments are in fact designed to be asymmetrical. In some, the specimen is placed in one arm of the instrument, such as in front of mirror M_2. Their use is concerned with 'amplitude spectroscopy', involving phase as well as amplitude, and requiring the calculation of the complex Fourier transform.

With regard to Eqn (7.50) it should be noted that it is assumed that the $l = 0$ position can be found accurately, and that the integration can be performed from $l = 0$ to $l = \infty$. The first of these is difficult and the latter impossible (Fig. 7.09). The more complicated methods mentioned above certainly help to overcome the first, and mathematically **apodization** is used as a means of minimizing the errors due to false termination of the Fourier transform integration (cf. the use of apodization in damping out the fringes around the Airy disc (§2.3)). However, the resolving power of the instrument, though limited, can clearly be very large since it is essentially determined by the upper limit of l available experimentally.

The interpretation of what is involved in this method can also understood in terms of autocorrelation. We have just noted that the introduction of the path difference l amounts to a time delay — a time delay between the arrival of the two wavetrains that are brought together to interfere. This is the very essence of autocorrelation (§4.7), since each pair of wavetrains originates from a single wavetrain before it has encountered the beamsplitter; in effect, every initial wavetrain is compared with itself further along its length. The time delay, τ say, is given by l/c where c is the velocity of the light. The amplitudes of each pair of recombining wavetrains can therefore be written as $A(t)$ and $A(t+\tau)$. What is measured at each value of l then, is $<A(t)A(t+\tau)>$. This is the autocorrelation function of the light, and we know from §7.4 that its Fourier transform is the power spectrum of the light, $|F(v)|^2$, (the Wiener–Khintchine theorem).

To return to the measurement of l. Alteration of l is achieved by the movement of a mirror, and the mechanical precision with which that can be controlled contributes towards setting a limit to the overall accuracy of the instrument. As a consequence, the method is particularly appropriate for work at long wavelengths, and it has been used extensively for some years in the infrared. However, with improvements in all aspects of the instrumentation the wavelength range has been extended down into the visible and even the ultraviolet.

Studies of the absorption by materials in the infrared are particularly important because of the type of information obtained. Materials absorb in that region for reasons associated with a number of phenomena. These include the vibrations of molecules (intra- and inter-molecular) and the rotations that occur about some chemical bonds.

Apart from their use in fundamental studies of materials, infrared spectra provide a powerful method of analysis. Figure 7.10 shows an

Fig. 7.10 — Spectra of light from Venus.
(a) Conventional grating spectroscopy.
(b) Fourier transform spectroscopy, showing detail of the rotational structure of the CO_2 band around $6505\,cm^{-1}$.
(After Connes, 1971.)

example that is interesting from several points of view. In (b) we have the spectrum obtained by Fourier spectroscopy, of the light from Venus (and therefore something of a link with the next section). It reveals the absorption of light by carbon dioxide molecules in the atmosphere of the planet: all the details of the spectrum can be identified with the vibrational modes of CO_2. In (a) we have the spectrum obtained with a conventional grating-spectrometer.

Then again, the atomic groups attached to the carbon-atom skeletons of organic molecules are usually the sites of the chemical reactivity of the molecule. These 'functional groups' have characteristic vibration frequencies, and from an analytical point of view they provide a simple, rapid and reliable 'fingerprint' for assigning a compound to its chemical class.

The technique has complementary uses in inorganic chemistry, including studies of the solid-state physical properties of materials. It is also applicable to the investigation of plasma effects, relaxation phenomena, etc.

For absorption studies of materials in the laboratory the light-source is matched to the spectral region to be investigated, and the material whose absorption spectrum is required is placed in a specimen cell at P in Fig. 7.08. As a beam-splitter for the far-infrared (wavenumber below c. $400 \, \text{cm}^{-1}$ — see Appendix K) a sheet of Mylar or polyethylene may be used. For measurements between $200 \, \text{cm}^{-1}$ and $4000 \, \text{cm}^{-1}$ devices consisting of germanium films deposited on various substrates (e.g. KBr, CsI) have been developed. And in the region from $2000 \, \text{cm}^{-1}$ to $16 \, 000 \, \text{cm}^{-1}$ films of ferric oxide, Fe_2O_3, on a substrate such as CaF_2, are employed. A compensator plate (not shown) may be used.

The choice of detector again depends on the spectral region. A Golay cell, pyroelectric bolometer, or a detector of the semiconductor photoconducting type are typically used.

To minimize absorption by atmospheric water vapour the whole system is evacuated to 10^{-2} to 10^{-3} Torr. Even so, it is of course necessary to obtain interferograms with and without the specimen present, to correct for the characteristic interferogram of the instrument.

All spectroscopic work in the infrared has one feature in common: the low energy associated with the long wavelengths involved. In the far-infrared, where wavelengths extend into the radio microwave range, the energy of the radiation can be more than a thousand times lower than in the visible spectrum (cf. Appendix K). This is where, as we have seen, the interferometry method scores over the classical methods of spectroscopy which employ prism and grating spectrometers. There are two reasons for this. Firstly, as P. Jacquinot showed, all interferometers that have circular symmetry, such as the Michelson, accept a larger light flux than slit spectrometers of equal resolving power (Jacquinot and Dufour, 1948; see also Jacquinot, 1960). Secondly, the interferometry method 'multiplexes' a spectrum, by which is meant that it yields information about the whole of a spectrum simultaneously: this was put into quantitative terms by P. Fellgett in 1951 (see also Fellgett, 1958). During a traverse of the moveable mirror all the light entering the interferometer is used in the production of the interferogram and the latter contains all the information about the spectrum of the light. As a consequence, there is a considerable increase in the signal-to-noise ratio over that experienced with a dispersive instrument such as a grating spectrometer, where a spectrum is scanned wavelength by wavelength. (With some types of detector the signal-to-noise ratio can be further improved by signal-averaging techniques, and these are now commonly employed.)

In addition to spectroscopy in the infrared, and in the visible and ultraviolet, mentioned earlier, Fourier transform methods are now employed in other types of spectroscopy, including nuclear magnetic resonance (NMR) spectroscopy, mass spectroscopy and a form of the latter known as ion cyclotron resonance (ICR) spectroscopy.

7.6 APPLICATIONS IN ASTRONOMY

A glance at the expression

$$\theta = \frac{1.22\lambda}{a} \qquad \{2.07\}$$

for the **angular resolution limit,** θ, of a telescope of aperture diameter a, shows that in the radio spectrum of wavelengths in the range $10\,\text{cm}$–$10^5\,\text{cm}$ (Appendix K), a telescope reflector would need to have diameter 2.10^5–2.10^9 times larger than that of the equivalent optical telescope working at a wavelength of about $500\,\text{nm}$. Added to that is the fact that whereas an optical telescope forms an image of stars in its field of view, a radio telescope gives only a single signal in the form of a voltage resulting from the radiation it receives over the entire range of directions determined by its resolving power. Small wonder that for some years after cosmic or galactic noise was discovered by K. G. Jansky in 1931 there was uncertainty as to where it was coming from.

Various observations were made by Jansky and others in the years that followed. By 1945 it was reported by E. V. Appleton that some radio emissions in the $60\,\text{mHz}$ ($5\,\text{m}$) band were associated with strong sunspot activity. In 1946 J. S. Hey and his co-workers discovered variations in the intensity of galactic noise from the direction of the constellation of Cygnus, suggesting that this particular radiation had its origins in a discrete source outside the solar system. So began the exploration of the radio sky.

This is certainly not the place to attempt to document the fascinating and spectacular history of radio astronomy. Nevertheless, notes on two of the important developments involving the use of Fourier methods are appropriate in this little book.

7.6.1 Aperture synthesis

In 1946 radar aerials had an angular beamwidth of about $10°$ at a wavelength of $1.5\,\text{m}$, which was quite inadequate for isolating, for example, regions on the surface of the sun from the general galactic background. Interferometry was a way out of this difficulty and J. L. Pawsey and his co-workers in Australia, making similar observations to Appleton's, used an aerial located high on a cliff overlooking the sea at Sydney. This arrangement (Fig. 7.11) formed an interferometer analogous to Lloyd's mirror experiment in optics. Interference occurred between direct radio signals and their reflection at the surface of the sea. The method was essentially the same as that involved in using a Michelson stellar interferometer but with the disadvantage of having a fixed baseline. Using the same aerial, J. Bolton and G. Stanley (1948) successfully recorded fringes produced by the Cygnus source — a constellation that conveniently rises only just above the horizon in Sydney. The workers in Australia also found other sources including a small intense

Fig. 7.11 — The sea interferometer.

source in the constellation of Taurus. This, alone among the early 'radio stars', was rapidly identified — with the Crab Nebula.

At the same time M. Ryle and D. D. Vonberg, in Cambridge, were also studying radiation from the sun. They were using an interferometer in the form of a direct analogue of the Michelson (Fig. 7.12(a)). It had two aerial

Fig. 7.12 — Radio analogue of Michelson stellar interferometer.

systems (dipole type) horizontally separated in the E–W direction, with the combined incoming signals fed to receiver equipment. Figure 7.12(b) shows the type of 'reception polar diagram' of such an interferometer, the envelope being the polar diagram of one aerial alone. The reasoning behind this scheme was that if the separation between successive minima of this 'double aperture' arrangement is made sufficiently large compared with the angular diameter of the sun, then as the aerial polar diagram is swept across the sun by the Earth's rotation any radiation from the sun would be detected by an oscillatory trace, whilst that from the galactic background would not.

The accuracy offered by this method is appreciated by considering Fig. 7.12(c) where we note that the angular width of the lobes in (b) is one half the width of the central maximum of a parabolic aerial of diameter equal to the distance between the two aerials comprising the interferometer. To put it another way, the ability of the interferometer to locate sources and measure their angular diameters could only be matched by a conventional paraboloid radio telescope if it had a diameter equal to twice the separation of the two aerials used in the interferometer.

During the appearance of a large sunspot in 1946, when the solar radiation was considerably increased, Ryle and Vonberg used their apparatus to determine the angular diameter of the radio source in the sun. For different separations of the aerials they measured the ratio of the maximum to the minimum in the fringes comprising the oscillatory traces. From their results they deduced that the source had a diameter of 10 minutes of arc. As this was not appreciably greater than the diameter of the visually seen sunspot they deduced that the radio source was related to the visual spot, or at least associated with it.

Using the two-aerial interferometer, Ryle and F. G. Smith (1948) discovered the strongest source in the northern sky, in Cassiopeia, but without any optical identification. They estimated that it has an angular diameter of less than 6 minutes of arc.

The Fourier aspect of variable-baseline interferometry for the derivation of the actual structure of a radio source was recognized by Pawsey and his collaborators in their work referred to above. It was used by H. M. Stanier in Cambridge in an investigation reported in 1950, to test theories developed in the late 1940s that emission from the sun in the absence of sunspots was particularly strong from the 'limb' of the sun at wavelengths of about 60 cm. In essentially the same way as that described in connection with the Michelson interferometer (§7.2.2) the fringe visibility was measured for aerial spacings up to 365 wavelengths. As the orientation of the aerial system was fixed the calculations had to be based on the assumption of circular symmetry in the source. Fourier transformation of the visibility curve gave a radial intensity distribution. (Strictly, this would be a Fourier–Bessel transformation.) Figure 7.13 shows the general form of the results, with no evidence of the limb-brightening that some investigators had expected.

The correlation aspect of the interferometer methods plays a very important role to which we now turn.

In Fig. 7.12(a) υ_1 and υ_2 are the voltages that are detected at the two

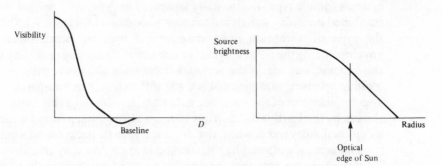

Fig. 7.13 — Visibility curve and radial distribution of radio 'brightness' across the
solar disc. (Based on Stanier, 1950.)

aerials and fed to a receiver whose output is proportional to the average of
their product. Now v_1 and v_2 each contain a 'noise' component because the
individual aerials have a large beamwidth. By its extraneous nature this
noise is uncorrelated between 1 and 2. We can therefore write the time-
average of the product of v_1 and v_2

$$\langle v_1 v_2 \rangle = \langle (v_{u1} + v_{c1})(v_{u2} + v_{c2}) \rangle$$
$$= \langle v_{u1} v_{u2} \rangle + \langle v_{u1} v_{c2} \rangle$$
$$+ \langle v_{c1} v_{u2} \rangle + \langle v_{c1} v_{c2} \rangle \qquad (7.51)$$

where the subscripts u and c refer to uncorrelated and correlated signals
respectively.

If the averaging time were sufficient all terms except the last one would
be zero since they are products involving uncorrelated voltages. The time-
averaged product of v_1 and v_2 would simply be

$$\langle v_1 v_2 \rangle = \langle v_{c1} v_{c2} \rangle . \qquad (7.52)$$

However, the averaging time has to be finite, noise fluctuations do persist
and the wanted signal is liable to be small compared with unwanted noise.
To minimize this, Ryle (1952) introduced a phase-switching procedure in
which the aerials were coupled alternately in and out of phase by means of a
$\frac{1}{2}\lambda$ length of transmission line in the cable from one aerial (Fig. 7.12(a)). The
difference between the two outputs was measured, giving

$$\langle (v_1 + v_2)^2 \rangle - \langle (v_1 - v_2)^2 \rangle = 4 \langle v_1 v_2 \rangle . \qquad (7.53)$$

More elaborate methods have subsequently been devised but the principle remains the same.

Expressed in this way the interferometer is measuring the cross-correlation between the signals at the two aerials as a function of the aerial separation; its Fourier transform is the spatial brightness distribution of the source (§7.4.1).

A number of different **unfilled aperture** systems were developed for the study of the distribution of the galactic emission in order to improve angular resolution. It was only the ability to measure the cross-correlation (i.e. $< v_{c1} v_{c2} >$) which allowed the polar-diagrams of two dissimilar aerials to be 'multiplied'. And it is *this* principle, whereby the effective area for 2-dimensional angular resolution no longer has to be the same as the area for sensitivity, that is implied by the term 'unfilled aperture'.

Ryle (1952) described the advantage of long arrays and various phase-switching schemes to improve performance. A notable example has been the 'Mills Cross' in Australia, named after its designer B. Y. Mills. Various forms of this interferometer were subsequently built in several countries. The original version (Mills and Little, 1953) consisted of two mutually perpendicular thin 'apertures' in the form of linear arrays running N–S and E–W (Fig. 7.14(a)). Each array comprised 250 dipoles (at 3.5 m wavelength)

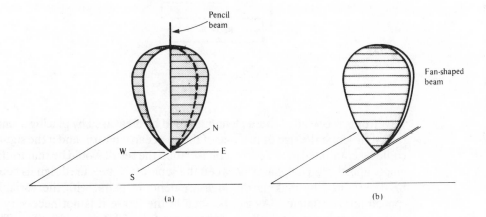

(a)

(b)

Fig. 7.14 — The Mills Cross.

extending over a length of 1500 ft and consequently had a polar diagram rather like a fan, as shown in (b). Combined together, the region of interaction of the two fans formed a narrow 'pencil' beam and, by an ingenious method of switching, only the signals received in this narrow beam were measured. To alter the elevation of the beam, phase differences were introduced between all the elements in the N–S array by switiching in the appropriate lengths of cable.

So far we have outlined in very general terms some of the main optical aspects of achieving directional sensitivity at radio wavelengths, and how the 'unfilled aperture' methods succeed in that respect. But their collecting areas are relatively small and this makes them unsuitable for investigations requiring great sensitivity — where the critical factor is the smallest radio flux that can be detected with a system. In connection with the complex problem of making surveys of distant radio 'stars', in which the location of weak signals is involved, a team led by Ryle at the Mullard Radio Astronomy Observatory of the Cavendish Laboratory at Cambridge pioneered the use of the variable spacing interferometer for 'aperture synthesis'. The interferometer was used in such a way as to achieve not only the angular resolution of a single very large aerial but also to go some way to matching it in sensitivity too. The method, elaborated in a paper by Ryle and Hewish in 1960, can be visualized by thinking of the single large aperture that is ideally required, as divided into separate elements of area (Fig. 7.15). If the

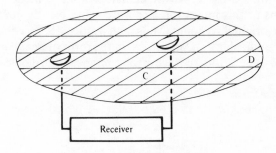

Fig. 7.15 — Aperture synthesis.

wavefronts received at these elements could be measured by placing a small radio telescope the size of just one of them, at each in turn; and if the signals could be combined in the correct phase; then the result would be that for the whole aperture. The phases between the separately measured signals could be determined by making the measurements in pairs, with the variable-spaced interferometer. As can be seen in the figure it is not necessary to make measurements for all the pairs; some (e.g. CD) are redundant. This type of synthesis was first accomplished in 1957 by Ryle's colleague J. H. Blythe, at a wavelength of 7.9 m and an effective angular resolution of approximately 1°.

Without the need for the two elements of the interferometer used in aperture synthesis to be equal in size and shape, various systems have evolved, for example employing a large fixed array in conjunction with a small, steerable parabolic aerial.

By utilizing the Earth's rotation a method of '**supersynthesis**' was initiated by Ryle in 1962. The principle is indicated in Fig. 7.16 where the two aerials A and B of an interferometer are separated by distance D along the E–W direction. The baseline rotates with the Earth every 24 hours and in one rotation each aerial has occupied all positions on an annular ring of radius D relative to the other. By using different values of D a complete circular area is synthesized, of radius equal to the maximum separation of

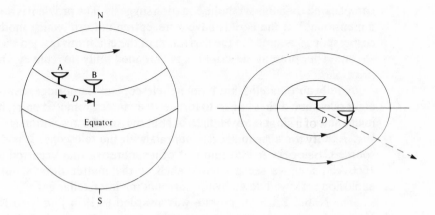

Fig. 7.16 — Supersynthesis.

the two aerials. (Strictly, the areas are ellipses except in the direction of the Pole star.)

In these synthesis methods the receiver circuitry, schematically shown in Fig. 7.17, allows the cosine and sine components, $C(D)$ and $S(D)$, of the

Fig. 7.17 — Sine and cosine components provide phase data.

interferometer receiver output to be measured for each value of D and projected angle of baseline. Reference to Eqns (7.12) and (7.13) will show that in the general case, and using exponential notation, we can write

$$C(D) + iS(D) = \int_{\phi} B(\phi)\, e^{2\pi i D\phi/\lambda}\, d\phi \qquad (7.54)$$

so that by Fourier transformation the brightness distribution $B(\phi)$ can be calculated. In practice it is of course theoretically and technically more complicated than this symbolic equation suggests. The problem is not only 3-dimensional but the aerials have a 'reception pattern' which modifies the $\cos^2$ response assumed in the derivation of the equations quoted above; and so on. The present description is intended only to convey the main principles.

A fully automated 5-km baseline telescope at Cambridge, consisting of eight telescope dishes linked to form sixteen interferometric pairs, has been in use since 1972: at a wavelength of 6 cm the angular resolution is 2 arc sec. Fig. 7.18 shows a radio source mapped with the telescope. It is the radio source Cassiopeia A, the remnant of a supernova that occurred in about 1660 AD, and we see a chaotic shell as the matter thrown out by the explosion is slowed down by the surrounding interstellar gas.

The Nobel Prize for physics was awarded in 1974 jointly to Ryle and Hewish for their pioneering research in radio astrophysics, Ryle for his observations and inventions, in particular the technique of aperture synthesis, and Hewish for the decisive role he played in the discovery of pulsars (— these were discovered with a different type of aerial, a massive array of 2048 dipoles covering an area of 18 km^2, that had been specially designed for the study of rapidly 'twinkling' radio sources for which aperture synthesis by its nature is unsuitable).

The principal telescope today is the **very large array** (VLA) near Soccoro, New Mexico, which has 27 antennas distributed in a Y configuration covering some 30 km. With the aid of such instruments the radio sources of intermediate angular size have now been mapped in detail.

Many radio sources, however, in particular the quasi-stellar objects, are still too small to be resolved. This prompted attempts to improve resolution still further by extending the baselines used. Large dish-type aerials, of which there are now many of different shapes and sizes around the world, have been used in the development of **very long-baseline interferometry** (VLBI). In such long-baseline work signals at each aerial can be recorded separately on magnetic tape, then brought together and replayed with great timing accuracy, made possible by the use of modern atomic clocks, to produce the required interferometer output. But phase variations also arise due to changing atmospheric conditions over the individual telescopes, having an effect that is equivalent to the positions of interference fringe maxima being fuzzy. The limitation imposed by this was overcome to a considerable extent by a new method. Dormant since the basis of it was conceived in 1958 by R. C. Jennison at Jodrell bank, it was realized afresh and applied to long-baseline interferometry by A. E. Rogers in 1974 in the USA. This **phase closure** method is based on the fact that if phases are measured between pairs of aerials that form a closed loop, e.g. three in a triangle, then when they are added together the spurious effects cancel.

Originally planned in 1973, the Jodrell Bank multi-element radio-link interferometer network (MERLIN) was commissioned in 1981 using two 'home' telescopes (one being the 250 ft dish) and five 'outstation' telescopes located in other parts of Great Britain. Recently, a European network, in which the Jodrell Bank telescope participates, has been set up. And with

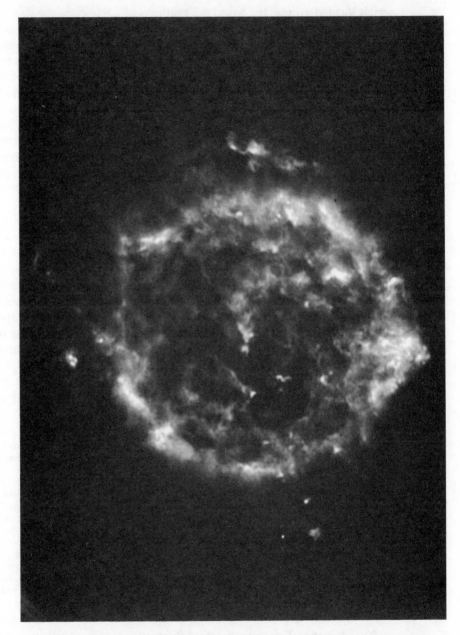

Fig. 7.18 — Radio photograph of the supernova remnant Cassiopeia A. The explosion of a massive star about 300 years ago has left a fast-moving ($7000\,\mathrm{km\,s^{-1}}$) shell of ejecta that is now interacting with the surrounding interstellar medium. Data for this map were collected by the 5-kilometre radio interferometer at Cambridge, operating at a frequency of 5 GHz. The angular diameter of the remnant is 5 arc min. and the smallest features visible have an angular size of 1 arc sec. It has been processed using the maximum entropy method (Appendix H), which is able to utilize the full resolution of the telescope, yet without introducing Fourier truncation sidelobes or other artefacts. (Courtesy of Dr S. F. Gull, Mullard Radio Astronomy Observatory, Cambridge.)

aerials located around the world the synthesis of a telescope the size of the earth has become a reality, giving 'images' that are unaffected by atmospheric and other effects, and with a resolution of 0.001 arc sec, which is approximately the angular diameter of the hot, visible stars. This is better than can be achieved with the best ground-based conventional optical telescopes because the images they give are severely blurred by the effects of the turbulence in the Earth's atmosphere.

To answer the most challenging questions, however, such as the nature of the 'central engine' at the heart of quasars and radio galaxies, believed to be a super-massive black hole, radio astronomy is going into space. A radio dish orbiting the Earth (QUASAT) as an outrigger station for the ground-based telescopes will provide an interferometer baseline of 20000 km, relaying its data to ground stations to be recorded on magnetic tape.

7.6.2 The intensity interferometer
The correlation aspect of the interferometry methods led to another valuable development in astronomy, both radio and optical, to which we must now turn. It is one which appeared at first to contradict the requirements for observing interference effects: it combines intensities, not amplitudes.

The principle of the intensity interferometer is due to R. Hanbury Brown who at Jodrell Bank in 1949 was considering two problems confronting radio astronomers: how to work with really long baselines, and how to overcome the effects of atmospheric turbulence.

In his book *The Intensity Interferometer* Hanbury Brown recalls how the challenge in 1949 was to measure the angular sizes of the two most prominent radio sources in the sky, Cygnus A and Cassiopeia A:

> At that time we knew only that their angular sizes could not be much greater than a few minutes of arc but we had no evidence as to how small they might be. If, as some people thought, they proved to be as small as the visible stars, then at metre wavelengths we should need two stations at the ends of the Earth. Was it feasible to build a radio interferometer with a baseline that could be extended if necessary from tens to hundreds, or perhaps thousands, of kilometres? The immediate technical difficulty in adapting conventional designs was to provide a coherent oscillator at the two distant points, and I started to wonder if this was really necessary. Could one perhaps compare the radio waves recieved at two points by some other means? As an example, I imagined a simple detector which demodulated waves from the source and displayed them as the usual noise which one sees on a cathode-ray oscilloscope. If one could take simultaneous photographs of the noise at two stations, would the two pictures look the same? This question led directly to the idea of the correlation of intensity fluctuations and to the principle of intensity interferometry.

The idea, therefore, was that if the fluctuations in intensity at two closely separated aerials correlated then the decrease in correlation (hence '**correlation interferometer**') as the baseline is increased would allow the angular diameter of the source to be determined (it would be the intensity analogue of Michelson's method for measuring the diameters of optically visible stars). The problem of the mutual instability of widely separated oscillators would be overcome. (At that time the atomic clocks now used in long-baseline interferometry had not been developed.)

The idea was successfully tried at Jodrell Bank (Hanbury Brown and Jennison, 1950), with an interferometer operating at a frequency of 125 MHz, to measure the angular diameter of the Sun. A year later it was used to measure the two radio sources that had posed the challenge. In the event they proved to be larger than expected and required only a baseline extension to a few kilometres. Though feeling that a sledgehammer had been used to crack a nut, Hanbury Brown and his collaborators found that when the radio sources were scintillating violently, due to ionospheric irregularities, the measurements of correlation were not significantly affected. The other problem had therefore been solved too.

The theory of the **intensity radio-interferometer** was put on a formal basis in 1954 by Hanbury Brown and R. Q. Twiss. Though of limited use in radio astronomy, because it is essentially a method in which the signal being measured must be strong compared with the noise level of the receivers, the important outcome has been in **optical astronomy**. It was the solution of the problem concerning atmospheric turbulence that prompted an enquiry into whether the intensity interferometer could usefully be employed in optical astronomy. Atmospheric turbulence†, it will be recalled, was a serious handicap in Michelson's method for measuring stellar diameters.

In 1956 a version of the instrument (Fig. 7.19) using two searchlight

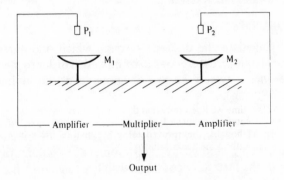

Fig. 7.19 — The intensity (or 'correlation') interferometer.

mirrors (M_1 and M_2) of 1.56 m diameter and baseline variable up to 14 m was used to measure the angular diameter of Sirius for the first time. Individually the mirrors produced blurred images of 8 mm diameter which were directed on to the cathodes of photomultipliers (P_1 and P_2). After amplification the signals were multiplied together. The average value of the product over some hours gave a direct measure of the cross-correlation between the intensity fluctuations in the light at the two mirrors. The angular diameter of Sirius deduced from the loss of correlation with increasing baseline (cf. the

---

† For general-purpose optical astronomy the problems due to our atmosphere are substantially reduced in the specification of the **Space Telescope**. With a 2.4-m diameter mirror orbiting at an altitude of 500 km the maximum spatial resolution will be of the order of 0.1 arc sec, compared with ground-based telescopes which cannot better approximately 1 arc sec.

loss of visibility in the Michelson method) was 0.0068 ± 0.0005 sec of arc, in good agreement with a value of 0.0063 ± 0.0006 sec of arc predicted by astrophysical theory on entirely different grounds.

The theory of the **optical intensity-interferometer** (Hanbury Brown andTwiss, 1958) is complicated by the quantum nature of the photoelectric effect.

In Fig. 7.19 let the instantaneous intensities at M_1 and M_2 be $I_1(t)$ and$I_2(t)$ respectively. The fluctuation in $I_1(t)$ can be expressed as $\Delta I_1(t) = I_1(t) - <I_1(t)>$ and similarly for I_2. Assuming the output currents from the photomultipliers to be proportional to the intensity of the light incident on them, the time-averaged product of the currents from the two photomultipliers is therefore proportional to $<\Delta I_1 \Delta I_2>$. Statistical calculations have shown that the latter is proportional to the square of the modulus of the cross-correlation between M_1 and M_2. Fourier transformation therefore gives the spatial brightness distribution of the source if, as is justifiable when measuring stellar diameters, it is assumed that the source is symmetrical.

Relatively insensitive to imperfections in mirrors and instability in the baseline, and less troubled than its forerunner with atmospheric disturbances, this has become an important technique in optical astronomy and has been actively pursued in more recent years by Hanbury Brown and his collaborators in Australia.

PROBLEMS

7.1 Calculate the distance through which one mirror of the Michelson spectral interferometer has to be moved in order to measure the wavelength difference between the two components of the sodium doublet ($589.0\,$nm and $589.6\,$nm).

7.2 In §7.2.2, when the angular brightness distribution, $B(\phi)$, of a radio source being studied by interferometry is symmetrical about its centre, $\phi = 0$, the interference fringe visibility function $V(D)$ is expressed in Eqn (7.12) as

$$V(D) = \frac{\displaystyle\int B(\phi) \cos{(2\pi D\phi/\lambda)}\, d\phi}{\displaystyle\int B(\phi)\, d\phi} .$$

Show that when $B(\phi)$ is constant over the entire width, ϕ, of the source, (i.e. from $\phi = -\phi_0/2$ to $+\phi_0/2$), then $V(D)$ is as given in Eqn (7.11), viz.

$$V(D) = \frac{\sin\dfrac{\pi\phi_0 D}{\lambda}}{\dfrac{\pi\phi_0 D}{\lambda}} .$$

Appendices

APPENDIX A — THE SCALAR-WAVE DESCRIPTION OF ELECTRO-MAGNETIC WAVES

A.1   The scalar-wave description

In Fig. A.01(a), if P rotates about C with a uniform speed then its projection

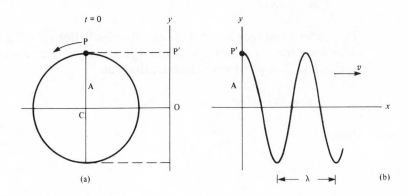

Fig. A.01

P′ on Oy describes simple harmonic oscillations about O. If P′ is attached to the end of a string, as illustrated in (b), the wavemotion transmitted along the string is analogous to the scalar-wave model for light (ignoring the question of polarization). To express this algebraically, we proceed as follows. First note that the oscillating displacement of P′ about O is given by

$$y = A \cos \frac{2\pi t}{T}$$

$$\left. = A \cos 2\pi \nu t \atop = A \cos \omega t \right\} \qquad (A.01)$$

where A = amplitude of oscillation
        T = period of oscillation (seconds)
        ν = frequency (oscillations/sec)
        ω = angular frequency (rads/sec) .

The equation for the wavemotion transmitted along the string may then be written in the form

$$y = A \cos \left[\omega t - \frac{2\pi x}{\lambda} \right] . \qquad (A.02)$$

The wave motion moves a distance of one wavelength (λ) for each oscillation and the velocity of the wave is therefore

$$\upsilon = \lambda / T$$

whence the usual relationship

$$\upsilon = \lambda \nu .$$

(The reader should confirm that the negative sign in Eqn (A.02) is associated with a wave travelling in the positive x direction, and vice versa.)
Eqn (A.02) is often written alternatively as

$$y = A \cos (\omega t - kx) \qquad (A.03)$$

where

$$k = 2\pi / \lambda$$

and therefore

$$\upsilon = \omega / k .$$

k is variously known as the **angular wavenumber, wavelength constant** or **wave constant**, and **propagation constant**. Often it is used to denote simply $1/\lambda$ and it is then called the (spectroscopic) wavenumber (the more conventional symbol, σ, for this is used in §6.3.2).
The wave considered so far has $y = A$ at $x = 0$ when $t = 0$. A similar

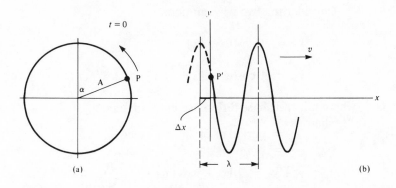

Fig. A.02

wave is illustrated in Fig. A.02(b) but it is delayed relative to the one we have just considered by distance Δx. Its equation is therefore

$$y = A \cos\left[\omega t - \frac{2\pi}{\lambda}(x + \Delta x)\right].$$

From Fig. A.02 it is evident that $\Delta x/\lambda = \alpha/2\pi$, and therefore

$$y = A \cos\left[\left(\omega t - \frac{2\pi x}{\lambda}\right) - \alpha\right]$$

$$= A \cos\left(\omega t - kx - \alpha\right) \tag{A.04}$$

where $\quad \alpha = \dfrac{2\pi}{\lambda}\Delta x$. $\tag{A.05}$

α is known as the **phase**, **phase angle**, or **phase constant**.

Relative to the wave represented by Eqn (A.03), α, is referred to as the **phase difference** and Δx as the **path difference**. Thus we have the much used expression

$$\text{phase difference} = \frac{2\pi}{\lambda} \times \text{path difference} .$$

A.2   Combining waves

A.2.1   *The principle of superposition: interference*

Often in this book it is necessary to find the resultant illumination when a number of light waves, differing only in amplitude and phase, overlap at some point — on a screen, say. According to the principle of superposition the resultant 'disturbance' at that point is the sum of the separate disturbances.

Consider the two wavemotions

$$y_1 = A \cos(\omega t - kx)$$

$$y_2 = A \cos(\omega t - kx - \alpha)$$

where α is the phase difference between them.
Then

$$y_1 + y_2 = 2A \cos\frac{\alpha}{2} \cos\left(\omega t - kx - \frac{\alpha}{2}\right)$$

which is of the form

$$Y = R \cos(\omega t - kx - \theta) \ . \tag{A.06}$$

This resultant is a light wave just like the separate incident waves, with frequency and velocity unaltered; but its amplitude, $2A \cos \alpha/2$, is not just the sum of the amplitudes of the separate incident waves. There is *interference* between the two, to an extent that depends on α. If the phase difference, α, between the two incident wavemotions is given by $\alpha = n2\pi$ (where n is zero or an integer), corresponding to zero path difference or a path difference of a whole number of wavelengths, the resultant does indeed have an amplitude of $2A$. However, if $\alpha = (n + \frac{1}{2})2\pi$ the two wavemotions are completely out of step and the resultant amplitude is zero. Phase differences between these two extremes give intermediate resultant amplitudes.

A similar result is obtained if the two incident amplitudes are unequal, though of course there could not be complete cancellation.

Note that if the above wavemotions separate again, beyond the place of intersection where they interfere, they proceed quite unaffected by the interference where they overlapped. Also note that *interference* can only occur between wavemotions having the same frequency.

For convenience, the place at which the effects of interference are to be calculated can be taken as $x = 0$. Eqn (A.04) becomes the displacement

$$y = A \cos(\omega t - \alpha) \ . \tag{A.07}$$

The resultant displacement due to wavemotions of different amplitudes and phases can be expressed as

$$Y = \sum_n A_n \cos(\omega t - \alpha_n) \tag{A.08}$$

which takes the form

$$Y = R\cos(\omega t - \theta) . \tag{A.09}$$

The summation is sometimes most easily performed with the aid of a phasor diagram as described in the following section, sometimes algebraically. For algebraic summation the use of exponential notation (§A2.3) is often simplest.

A.2.2 Phasor diagrams

Eqn (A.07) can be expanded as

$$y = A\cos\alpha \, \cos\omega t + A\sin\alpha \, \sin\omega t$$
$$= A_c\cos\omega t + A_s\sin\omega t$$

where $\quad A_c = A\cos\alpha, \quad A_s = A\sin\alpha$.

A_c and A_s then carry the information about the amplitude and phase of the wave, as illustrated in Fig. A.03(a). We have

$$A = (A_c^2 + A_s^2)^{\frac{1}{2}}$$

and

$$\alpha = \tan^{-1}(A_s/A_c) .$$

(a)

(b)

Fig. A.03

A_c and A_s are like the components of a two-dimensional vector but since angles here represent phase instead of direction the term *phasor* is used to make the distinction. The summation of Eqn (A.08) may therefore be

performed graphically and Fig. A.03(b) shows that the resultant amplitude
(R in Eqn (A.09)) is given by

$$R = \left[\left(\sum A_n \sin \alpha_n\right)^2 + \left(\sum A_n \cos \alpha_n\right)^2\right]^{\frac{1}{2}}$$

and the phase

$$\theta = \tan^{-1}\left[\frac{\sum A_n \sin \alpha_n}{\sum A_n \cos \alpha_n}\right].$$ (A.10)

A.2.3   *Exponential representation*

A commonly used alternative procedure is based on expressing

$$y = A \cos(\omega t - \alpha)$$

in the form

$$y = \Re A\, e^{i(\omega t - \alpha)}$$ (A.11)

where $\Re$ indicates that only the real part of the expression is intended: for
brevity $\Re$ is usually omitted, but we retain it here for clarity.

The summation (Eqn (A.08))

$$Y = \sum_n A_n \cos(\omega t - \alpha_n)$$

is then written as

$$Y = \sum_n \Re A_n\, e^{i(\omega t - \alpha_n)}$$ (A.12)

Now, when adding complex quantities, the real parts and the imaginary
parts are added separately. Therefore the sum of the real parts of all the
terms in this summation is equal to the real part of the whole summation. So
we can rewrite the above as follows

$$Y = \Re \sum_n A_n\, e^{i(\omega t - \alpha_n)}$$ (A.13)

As in §A.2.1, we know that the result of the summation is of the form (Eqn (A.09))

$$Y = R\cos(\omega t - \theta) \, ,$$

which we can now express as

$$Y = \mathcal{R}R \, e^{i(\omega t - \theta)} \, . \tag{A.14}$$

Equating Eqns (A.13) and (A.14), we have

$$\mathcal{R}R \, e^{i(\omega t - \theta)} = \mathcal{R}\sum_n A_n \, e^{i(\omega t - \alpha_n)} \, , \tag{A.15}$$

We can separate out the harmonic time factor $e^{i\omega t}$ by rewriting the above as

$$\mathcal{R}(R \, e^{-i\theta} . e^{i\omega t}) = \mathcal{R}\left[\left(\sum_n A_n \, e^{-i\alpha_n}\right) e^{i\omega t}\right] \tag{A.16}$$

where the first factor on each side is the **complex amplitude**. And it is the modulus of $\sum_n A_n \, e^{-i\alpha_n}$ that will therefore give us the value of the resultant amplitude, R, that we seek, for we have

$$R \, e^{-i\theta} = \sum_n A_n \, e^{-i\alpha_n} \, . \tag{A.17}$$

If the summation is performed algebraically (or as an integration in some instances), the result will be a complex quantity, Z say, and we obtain its modulus by multiplying it by its complex conjugate thus

$$Z . Z^* = |Z|^2 = R^2 \, . \tag{A.18}$$

The use of exponentials in the way we have described here is like the phasor method but with the summation in Eqn (A.17) performed in the complex plane. To appreciate this we write the complex amplitudes in Eqn (A.17) as

$$R\cos\theta - iR\sin\theta = \sum A_n\cos\alpha_n - i\sum A_n\sin\alpha_n \tag{A.19}$$

and equating real and imaginary parts gives

$$R \cos \theta = \sum A_n \cos \alpha_n$$

$$R \sin \theta = \sum A_n \sin \alpha_n$$

whence

$$R = \left[\left(\sum A_n \cos \alpha_n \right)^2 + \left(\sum A_n \sin \alpha_n \right)^2 \right]^{\frac{1}{2}} \qquad (A.20)$$

$$\theta = \tan^{-1} \left[\frac{\sum A_n \sin \alpha_n}{\sum A_n \cos \alpha_n} \right] \qquad (A.21)$$

as in Eqns (A.10).

Concerning the observation that a phasor diagram may be regarded as the addition of complex amplitudes in the complex plane, it should be noted that rotating a phasor through an angle α corresponds to multiplying the complex amplitude by $e^{i\alpha}$. It is because the latter type of operation is often mathematically easier than the manipulation of trigonometrical formulae that exponential notation is so often used. An example is to be found in the discussion of Eqn (5.13).

APPENDIX B — THE STOKES TREATMENT OF REFLECTION AND REFRACTION

This classical treatment runs as follows. In Fig. B01(a) a ray of light of

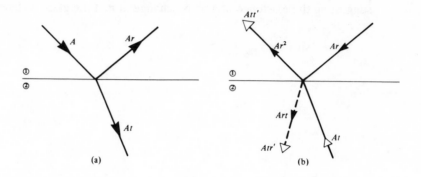

Fig. B.01

amplitude A gives a reflection of amplitude Ar and a refracted ray of amplitude At at the interface between two dielectric media, 1 and 2, of different refractive indices. r and t denote the fraction of the amplitude reflected and refracted (transmitted) respectively.

On the basis of the reversibility of light rays, if these reflected and refracted rays are reversed they should amount to a reversal of the incident ray of amplitude A. This is shown schematically in (b). The reversal of the ray of amplitude Ar gives a reflection back into the first medium of amplitude Ar^2, and a refracted ray of amplitude Art. Similarly, the reversal of the ray of amplitude At gives a reflection back into the second medium of amplitude Atr' and a transmitted ray into the first medium of amplitude Att', where r' and t' are fractions appropriate for events at the 2/1 interface as opposed to the 1/2 interface.

To reproduce (a) in reverse we therefore have to equate as follows

$$Att' + Ar^2 = A$$

i.e.
$$tt' = 1 - r^2 \tag{B.01}$$

and

$$Art + Atr' = 0$$

i.e.
$$r = -r' . \tag{B.02}$$

The difference in sign of the amplitudes in this last equation indicates a phase change of π in the reflection at either 1/2 or 2/1. Experimental observation shows that it normally occurs when light is incident at a boundary from the side of higher velocity (lower refractive index). For example, in Lloyd's mirror experiment, where light reflected at a glass surface interferes with light not reflected, the interference fringe corresponding to no path difference (light at grazing incidence) is dark, not light, suggesting the existence of a phase change of π at the glass surface.

APPENDIX C — CIRCULAR-APERTURE FRAUNHOFER DIFFRACTION PATTERN

The derivation in §2.2 of the Fraunhofer diffraction pattern given by a uniformly illuminated slit is comparatively simple for two reasons: (1) the elemental strips into which we visualise the slit being divided (parallel to its length) all have the same area, and the amplitude of the light leaving each is consequently the same, (2) the phase of the illumination from successive strips varies linearly across the slit for any given off-axis direction. The expression (Eqn (2.05)) for the resultant from the whole width of the slit is readily obtained with the aid of a simple phasor diagram (Fig. 2.02), or by straightforward integration as described at the end of §4.2.

With a uniformly illuminated circular aperture the problem is more difficult because although the phase varies linearly across the aperture as for a slit, the areas of the successive elemental strips, and therefore the amplitudes of the light leaving them, are not all the same. From Fig. C.01 we

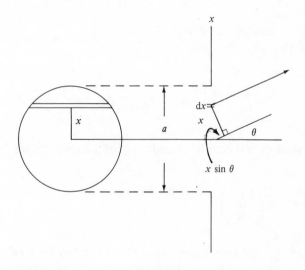

Fig. C.01

see that the phase for light leaving a strip at x is $(2\pi/\lambda)x\sin\theta = 2\pi ux$ as usual, but the area of the strip is $2(a^2/4 - x^2)^{\frac{1}{2}}\,\mathrm{d}x$. The phasor diagram would not now resemble part of a regular polygon; it would be equiangular but not equilateral. Algebraic integration offers the best approach. We can express the resultant amplitude, $F(u)$, as

$$F(u) = \int_{-a/2}^{+a/2} 2(a^2/4 - x^2)^{\frac{1}{2}}\, e^{i2\pi ux}\,\mathrm{d}x \; . \tag{C.01}$$

Though looking rather formidable this can fortunately be put into a standard integral form by making two substitutions: (1) replace the variable x by υ where

$$\upsilon = 2x/a ,\qquad\qquad\text{(C.02)}$$

which gives

$$dx = (a/2)\,d\upsilon ,$$

and (2) put

$$\pi u a = \rho ,$$

which is constant for each value of u.

Eqn (C.01) then becomes

$$F(u) = \frac{a^2}{2}\int_{-1}^{+1}(1-\upsilon^2)^{\frac{1}{2}}\,e^{i\rho\upsilon}\,d\upsilon .\qquad\qquad\text{(C.03)}$$

The integral here is a standard one, the result being

$$\pi\left[\frac{J_1(\rho)}{\rho}\right]$$

where J_1 is the **Bessel function** of the '**1st order, 1st kind**'.†

Though the term in [ ] brackets is much like the sinc function, it can be shown that it has a value of $\frac{1}{2}$ as $\rho \to 0$; so we introduce a factor of 2 for

---

†A Bessel function of **order n** is a solution of the differential equation

$$x^2\frac{d^2y}{dx^2}+x\frac{dy}{dx}+(x^2-n^2)y = 0 ,$$

this equation being known as **Bessel's equation of order** n.

The simplest solution for each order is the solution of the 'first kind' and is given by the Bessel function

$$J_n(x) = x^n\frac{(2n)!}{(n!2^n)^2}\left\{1-\frac{x^2}{2(2n+2)}+\frac{x^4}{2.4(2n+2)(2n+4)}-\cdots\right\}$$

These series expansions are rather like those for sine and cosine. Like the sinc function, $J_1(x)$ resembles a damped sine-wave.

making the comparison in Fig. 2.05. With $\rho = \pi u a$ we can write

$$F(u) = \frac{\pi a^2}{4}\left[\frac{2J_1(\pi u a)}{\pi u a}\right] . \tag{C.04}$$

Squared, this gives the expression describing the radial distribution of intensity in the Airy pattern shown in Fig. 2.04(a).

APPENDIX D — THE FOURIER TRANSFORM OF A LINEAR ARRAY OF N EQUALLY SPACED δ-FUNCTIONS

In §4.3 we noted that an array of δ-functions distributed along the x-axis at positions x_N is written as

$$f(x) = \sum_N \delta(x - x_N) \qquad \text{(4.21) (D.01)}$$

and that the Fourier transform of the array is

$$F(u) = \sum_N e^{2\pi i u x_N} . \qquad \text{(4.22) (D.02)}$$

As an example, we considered a pair of δ-functions at $x = \pm D/2$ and showed that the Fourier transform is the expression for the diffraction pattern given by two ideally narrow slits with the same separation as the δ-functions. We also showed that if the pair of δ-functions is moved with respect to the origin of x, then there is an overall change of phase through the diffraction pattern (the **shift theorem**).

In just the same way, the general case of a linear array of N δ-functions distance D apart gives the 'grating term' of Eqn (4.19). We will choose $N = 2M + 1$, so that we can have M δ-functions symmetrically either side of a central one at the origin. The Fourier transform of the array is then

$$F(u) = \sum_{n=-M}^{n=+M} e^{2\pi i u n D} . \qquad \text{(D.03)}$$

Expanding this gives

$$
\begin{aligned}
F(u) &= e^{2\pi i u(-M)D} + e^{2\pi i u(-M+1)D} + e^{2\pi i u(-M+2)D} + \ldots + e^{2\pi i u M D} \\
&= e^{2\pi i u(-M)D}(1 + e^{2\pi i u D} + e^{2\pi i u 2D} + \ldots + e^{2\pi i u 2MD}) .
\end{aligned}
\qquad \text{(D.04)}
$$

The series in brackets is a geometrical progression of $2M + 1$ terms. Therefore, we have

$$F(u) = e^{2\pi i u(-M)D}\left(\frac{1 - r^{2M+1}}{1 - r}\right) \qquad \text{(Appendix L) (D.05)}$$

where $r = e^{2\pi i u D}$, giving

$$F(u) = e^{-2\pi iuMD} \left(\frac{1 - e^{2\pi iu(2M+1)D}}{1 - e^{2\pi iuD}} \right) \tag{D.06}$$

$$= \frac{e^{-2\pi iuMD} \, e^{+2\pi iu[(2M+1)/2]D} \left(e^{-2\pi iu[(2M+1)/2]D} - e^{+2\pi iu[(2M+1)/2]D} \right)}{e^{+2\pi iu(D/2)} \left(e^{-2\pi iu(D/2)} - e^{+2\pi iu(D/2)} \right)}$$

$$= \frac{\sin \pi u(2M+1)D}{\sin \pi uD} = \frac{\sin N\pi uD}{\sin \pi uD} \ . \tag{4.19) (D.07}$$

APPENDIX E — DIFFRACTION OF X-RAYS BY CRYSTALS. THE EQUIVALENCE OF THE LAUE CONDITIONS AND THE BRAGG REFLECTION CONCEPT

The Laue conditions

The parallelepiped in Fig. E.01 depicts the unit cell of a crystal lattice, with

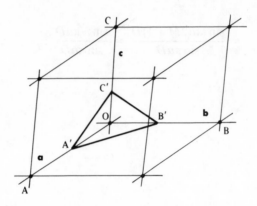

Fig. E.01

each lattice 'point' representing an identical group of atoms. In §2.7 it was noted that if a wavetrain of X-rays of wavelength λ sweeps through the crystal, then X-ray diffraction maxima occur in directions for which the scattering from the atomic group associated with each of the lattice points O, A, B and C is in phase. The Laue conditions (Eqn (2.18)) are a statement of this requirement, viz.

$$(\mathbf{s} - \mathbf{s}_0) \cdot \mathbf{a} = h\lambda$$
$$(\mathbf{s} - \mathbf{s}_0) \cdot \mathbf{b} = k\lambda \qquad\qquad (E.01)$$
$$(\mathbf{s} - \mathbf{s}_0) \cdot \mathbf{c} = l\lambda$$

where $\mathbf{s}_0$ is a unit vector defining the incident X-ray beam, and $\mathbf{s}$ is a unit vector defining the direction of a diffraction maximum; $\mathbf{a}, \mathbf{b}, \mathbf{c}$ are the lattice translation vectors (not necessarily orthogonal); and h, k, l are integers or zero.

Bragg reflection

In explaining his 'reflection' description of the same phenomenon, W. L. Bragg pointed out that the requirement that scattering from the atomic groups at A, B and C (we can call them **scattering centres**) be in phase is analogous to optical reflection in a mirror whose surface contains A, B, C: for any angle of incidence, reflection occurs at an angle equal to the angle of incidence. With this relation of reflected to incident wave directions the waves scattered from all scattering centres in the reflecting plane are in phase with each other, regardless of their positions within the plane — as in optical reflection (Fig. E.02).

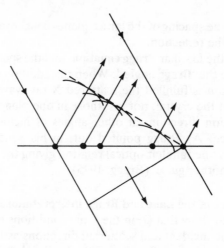

Fig. E.02 — The reflection of a wavefront by scattering centres lying in a plane (shown edge-on).

Unlike optical reflection, however, when X-rays are incident at a crystal lattice plane only a very small fraction of the amplitude of the incident beam is 'reflected'. Most of the radiation is transmitted through the crystal. Furthermore, reflections from the successive lattice planes parallel to the first will not, in general, be in phase with each other. Reinforcement can be obtained, however, by adjusting the angle of incidence. As shown in Fig. E.03, reinforcement requires that the path difference X′Y′Z–XYZ is a

Fig. E.03 — The reflection of a wavetrain by successive lattice-planes of spacing d.

whole number of wavelengths. This is equivalent to requiring that Y′Y–Y′V is a whole number of wavelengths or, since Y′Y = Y′W,

$$VW = n\lambda$$

i.e. $2d \sin \theta = n\lambda$ (E.02)

where d is the spacing of the lattice planes concerned, and the integer n is the 'order' of the reflection.

This is the familiar **Bragg equation**, and the special values of the glancing angle θ are the '**Bragg angles**'. When the additional condition expressed by this equation is fulfilled the scattered X-ray waves from all lattice points throughout the crystal, not just those in one plane, reinforce each other in the reflection direction: at other angles of incidence there is destructive interference. As Bragg pointed out, in his penetratingly clear way, the equation is 'the familiar optical relation giving the colours reflected by thin films, in another guise' (Bragg, 1975).

Equivalence of the Laue and Bragg interpretations

We need to show that (i) in the Laue conditions the unit vectors $\mathbf{s}_0$ and $\mathbf{s}$, defining the incident and scattering directions associated with a diffraction maximum, make equal angles to the lattice planes, defined by h, k, l, that can be thought of as reflecting in the Bragg sense, (ii) that the angle that $\mathbf{s}_0$ and s make to those planes, and the spacing of the planes (the perpendicular distance between them), are as given by the Bragg equation.

(i) In Fig. E.01 construct the plane A'B'C', such that $OA' = OA/h$, $OB' = OB/k, OC' = OC/l$, where h, k, l are as in the Laue equations. Since h, k, l are integers, or zero, it follows that A', B', C' lie in a lattice plane that passes through lattice points elsewhere in the same lattice (of which we have only drawn one unit cell). Therefore, the lattice plane through A'B'C' could 'reflect' X-rays in the Bragg sense, in some direction dependent on the direction of incidence and on the wavelength of the X-rays. (Note that if e.g. $h = 0$ then A' is at infinity, and this simply means that the lattice plane is parallel to the a-axis.)

We will now show that A'B'C', and therefore the other lattice planes parallel to it, are planes to which the s and $\mathbf{s}_0$ in the Laue conditions are equally inclined — as required for the Bragg reflection interpretation. A more convenient view of A'B'C' for this is shown in Fig. E.04(a).

Putting $\mathbf{s} - \mathbf{s}_0 = \mathbf{S}$, Eqns (E.01) become

$$\frac{\mathbf{S}}{\lambda} \cdot \frac{\mathbf{a}}{h} = 1$$

$$\frac{\mathbf{S}}{\lambda} \cdot \frac{\mathbf{b}}{k} = 1$$ (E.03)

$$\frac{\mathbf{S}}{\lambda} \cdot \frac{\mathbf{c}}{l} = 1 \ .$$

Subtracting the second from the first gives

$$\frac{\mathbf{S}}{\lambda} \cdot \left(\frac{\mathbf{a}}{h} - \frac{\mathbf{b}}{k}\right) = 0 \qquad (E.04)$$

i.e. $\mathbf{S}$ is perpendicular to $\overrightarrow{B'A'}$.

Subtracting the other pairs in Eqn (E.03) shows that $\mathbf{S}$ is also perpendicular to $\overrightarrow{A'C'}$ and $\overrightarrow{C'B'}$. $\mathbf{S}$ is therefore perpendicular to the plane $A'B'C'$.

Now in Fig. E.04(b) we see that $\mathbf{S}$ bisects $\mathbf{s}$ and $\mathbf{s}_0$. It follows, therefore, that

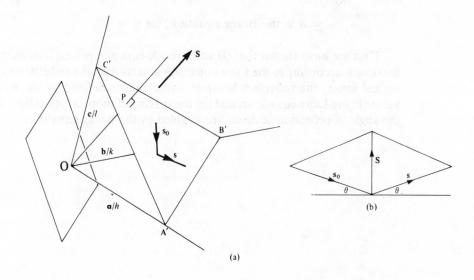

(a)

(b)

Fig. E.04

$\mathbf{s}$ and $\mathbf{s}_0$ are equally inclined to $A'B'C'$, in accord with the idea of reflection in the lattice plane $A'B'C'$. And by its construction this plane is defined by h, k, l in the Laue conditions.

(ii)   As points in the plane $A'B'C'$ scatter with a path difference of one wavelength compared with those in the parallel lattice-plane through O, we also have Bragg's picture of the reinforcement of scattering from successive lattice planes. If $\mathbf{s}_0$ and $\mathbf{s}$ are at the equal glancing angle to $A'B'C'$ that Bragg designated as θ, we should be able to show that the spacing, OP, of these planes is equal to the d in the Bragg equation, for $n = 1$. From the figure we have

$$OP = \frac{\mathbf{a}}{h} \cdot \frac{\mathbf{S}}{|\mathbf{S}|} . \qquad (E.05)$$

With the first of the Laue equations this gives

$$OP = \frac{\lambda}{|\mathbf{S}|} .$$ (E.06)

From Fig. E.04(b) we have $|\mathbf{S}| = 2\sin\theta$ and therefore

$$OP \doteq \frac{\lambda}{2\sin\theta}$$

(E.07)

$$= d \text{ in the Bragg equation, for } n = 1 .$$

Thus we have shown that (i) scattered X-rays comprising a diffraction maximum according to the Laue equations also constitute a reflection in the optical sense, the reflection being at lattice planes defined by the h, k, l values in the Laue equations, and (ii) the spacing of those lattice planes, and the angle of reflection at them, are as given by the Bragg equation.

APPENDIX F — SUMMARY NOTES ON CHAPTERS 1–5
F.1 Single-slit diffraction (Fig. 2.01)

Amplitude
$$R_1(u) = R_1(0)\frac{\sin \pi u a}{\pi u a} \tag{2.05}$$

where
$$u = \frac{\sin \theta}{\lambda} \tag{2.04}$$

and a is the slit width. (Derived by phasor diagram (§2.2) or by integration — see text related to Fig. 4.03.)

F.2 Grating diffraction (Fig. 2.10)

Amplitude pattern is the product of two terms, the 'single-aperture term' and the 'grating term':

$$R_N(u) = R_1(0)\left(\frac{\sin \pi u a}{\pi u a}\right)\left(\frac{\sin N\pi u D}{\sin \pi u D}\right). \tag{2.15}$$

Grating diffraction maxima (principal maxima), order $\pm n$, occur in directions for which there is reinforcement between the light diffracted from successive slits, i.e. when

$$D\sin \theta = n\lambda \qquad \text{(Fig. 2.10)}$$

i.e.
$$\frac{\sin \theta}{\lambda}(=u) = \frac{n}{D}. \tag{F.01}$$

In these special directions, the total amplitude from the grating is then simply the sum of the contributions, in those directions, from all the individual slits. Substitution of $u = n/D$ for those special directions, into Eqn (2.15) above, gives the amplitudes of the diffraction orders as

$$R_1(0)\left[\frac{\sin \dfrac{\pi n a}{D}}{\dfrac{\pi n a}{D}}\right]N, \tag{F.02}$$

cf. intensities Eqn (2.17).

Thus we regard (§2.5) the grating diffraction maxima as an **amplified sampling** of the single-slit pattern, the sampling being in directions given by $u = n/D$. ((F.02) can be derived directly from Eqn (2.05) above.)

This sampling aspect is not restricted to the case where the apertures comprising the grating are slits, though the sinc function is, of course, peculiar to that case. The concept is also directly extendable to two and three dimensions (§2.6 and §2.7 respectively).

F.3 The grating diffraction pattern as the Fourier analysis of the (overall) aperture function of a grating

In the case we are dealing with, where the overall aperture function, $f(x)$, of the grating is real and even, we can write the following cosine Fourier series to represent $f(x)$, assuming that the number of grating repeats is very large (strictly infinite):

$$f(x) = \sum_{-\infty}^{+\infty} \frac{A_n}{2} \cos \frac{2\pi nx}{D} \, , \qquad \{3.04\}$$

where the amplitudes, $A_n/2$, of the harmonics (the 'Fourier coefficients') are given by

$$\frac{A_n}{2} = \frac{1}{D} \int_D f(x) \cos \frac{2\pi nx}{D} \, dx \, , \qquad \{3.05\}$$

and n/D = spatial frequencies of the harmonics.

For the special case of the repeating rectangle function this gives

$$\frac{A_n}{2} = \frac{ha}{D} \left[\frac{\sin \dfrac{\pi na}{D}}{\dfrac{\pi na}{D}} \right] . \qquad \{3.06\}$$

These amplitudes of the harmonics comprising the Fourier analysis of the grating aperture function are thus proportional to the amplitudes of the diffraction maxima given by the grating (Eqn (F.02)). Furthermore, the directions of those diffraction maxima, expressed as $u = (\sin \theta)/\lambda$, give the spatial frequencies, n/D, of the harmonics (Eqn (F.01)).

The diffraction plane is therefore regarded as the **Fourier plane**, or **frequency domain** (§3.4.1). This is equally applicable to other types of grating, including 2- and 3-dimensional gratings.

F.4   Diffraction patterns as Fourier transforms

(a) *Single-aperture diffraction pattern*
When $f(x)$ is a non-periodic even function Eqns (3.04) and (3.05) become

$$F(x) = \int_{-\infty}^{+\infty} F(u) \cos 2\pi ux \, du \qquad \{4.05\}$$

and

$$F(u) = \int_{-\infty}^{+\infty} f(x) \cos 2\pi ux \, dx \, , \qquad \{4.06\}$$

where $f(x)$ and $F(u)$ are now the (**cosine**) **Fourier transform** of each other. (In the general case, Eqns (4.09) and (4.10) apply.)

If, for $f(x)$, we insert the details of the same rectangle function as before, the Fourier transform of $f(x)$ becomes

$$F(u) = ha\left(\frac{\sin \pi ua}{\pi ua}\right). \qquad \{4.07\}$$

This consists (ignoring the constant term) of the sinc function, just as in Eqn (2.05) above for the diffraction pattern given by a single slit having the same rectangle function for its aperture function. The diffraction pattern is therefore said to be the Fourier transform of the aperture function (and vice versa) (§4.2). This is again a statement that is applicable to other types of aperture.

(b) *Grating diffraction pattern*

(i) *As Fourier transform of the (overall) aperture function of a grating*
Since a grating diffraction pattern is an (amplified) sampling of the diffraction pattern given by one constituent repeat, D, of the grating alone (§F.2), it follows from (a) that the grating pattern is a sampling of the Fourier transform of the individual aperture function of the single unit D.

On the other hand, a grating is of finite extent, so its overall aperture function can be regarded as a single non-periodic unit (even though it contains a subunit (D) that repeats for a finite number of times). In that sense the grating diffraction pattern is described as the Fourier transform of what, for clarity, we have usually called the overall aperture function of the grating.

There is no conflict between these statements, and it is useful to be able to regard any diffraction pattern as the Fourier transform of the associated aperture function. We recognize this in §4.2 and use it in Chapter 5 (e.g. Fig. 5.06).

(ii) *As a product of transforms*

In the expression (Eqn (2.15) restated in F.2 above) for the diffraction pattern given by a grating, the 'grating term' is the Fourier transform of the array of δ-functions defining the distribution of apertures in the grating

(§4.3). As the 'single-aperture term' is the Fourier transform of the aperture function of the basic single repeat of the grating, it follows that the grating diffraction pattern can be regarded as the product of two Fourier transforms.

Furthermore, in §4.4.2 we show that the overall aperture function of a grating is the convolution of one single aperture function with the array of δ-functions that specify the distribution of that aperture function in the grating. And in (i), we have reminded ourselves that the grating pattern can be interpreted as the Fourier transform of its overall aperture function (§4.2). It follows, therefore, that the Fourier transform of the convolution of the single-aperture function and the array of δ-functions defining the grating is the product of their individual transforms. In other words, convolution in real space corresponds to multiplication in diffraction space (the frequency domain). This relationship is depicted schematically in the present context in Fig. 4.07. It is an example of the **convolution theorem** (§4.5).

F.5 Image formation

Model 1: image formation as a double Fourier process

This approach has its origins in the Abbe-Porter theory of image formation (§5.1). It considers the whole wavefront leaving an object, as opposed to the wavefront leaving each object point individually (Model 2).

In connection with the multiple-slit object grating in Fig. 5.04 we note that the amplitude of the nth order pair of diffraction maxima, in directions given by $u = \pm n/D$, are the Fourier coefficients of the overall aperture function of the grating (§3.4.1 and F.3). And in §5.3.1 we see that that pair of diffraction maxima contribute the sinusoidal illumination distribution of spatial frequency $\pm n/D$, to the formation of the image. The diffraction pattern is the Fourier analysis of the object grating, and the image is the Fourier synthesis of that part of the Fourier analysis allowed through the system by the aperture of the lens.

If the object is non-periodic the same applies in principle, though we are then dealing with two stages of Fourier transformation (i.e. integrals instead of series).

In the spirit of our remarks about gratings containing a finite number of apertures (F.4(b) (i) and §4.2) image formation can in general be described as a **double Fourier transformation process**.

Model 2: image formation as a convolution

This can be described as the Rayleigh–Duffieux model (§5.1). Each object point is regarded as a source of light and it is imaged as the Airy pattern of the aperture of the imaging lens (§1.3.2). The whole image distribution is then the convolution of the object distribution with this Airy pattern (or more complicated pattern if aberrations are present). The convolution is of complex amplitudes if the illumination is coherent, and intensities if incoherent: in the latter case the response of the imaging lens to the point object is the point spread function (PSF) of the system.

This model is valuable for the way it is related to electrical network theory and communications systems (Appendix I), and the notion of transfer functions (Fig. 5.02). The convolution theorem lies at the heart of this. Instead of calculating the convolution of an input with the 'impulse response' of a system, it is usually easier to calculate the transforms of the two, multiply them, and take the (then called) inverse transform of the product. (The transform of the PSF is the 'optical transfer function' (OTF).)

Incoherent processing is dealt with in §5.5.2.

Equivalence of the two models

The Abbe–Porter model can be expressed as

$$\text{Image} = T[T(\text{object}) \times (\text{Ap.fn})] \ . \tag{5.08}$$

Application of the convolution theorem to this gives

$$\text{Image} = \text{Object} \circledast T(\text{Ap.fn}) \tag{5.09}$$

which is a statement of the Rayleigh–Duffieux model.

APPENDIX G — THE LENGTH OF WAVETRAINS AND THEIR SPECTRAL WIDTH

Fig. G.01(a) represents a simple harmonic wave of frequency v_1 and infinite duration. In (b) it has been multiplied by a truncation function, here a rectangle function (shown dotted), so that it exists only for the time duration Δt. The truncated wave can be expressed as

$$f(t) = A \cos 2\pi v_1 t$$

$$= A e^{2\pi i v_1 t} \quad \text{(real part implied)} \tag{G.01}$$

for $|t| \leq \Delta t/2$ and $f(t) = 0$ for $|t| > \Delta t/2$.

One might expect the truncated wave still to be monochromatic, with frequency v_1, but we can readily show that this is not so.

A non-periodic function $f(t)$ can be represented by the summation of an infinite continuum of simple harmonics of different frequencies (the Fourier integral):

$$f(t) = \int_{-\infty}^{+\infty} F(v) \, e^{2\pi i v t} \, dv \tag{4.33}$$

where

$$F(v) = \int_{-\infty}^{+\infty} f(t) \, e^{-2\pi i v t} \, dt . \tag{4.34}$$

To apply this to the present example, we substitute (G.01) into {4.34}:

$$F(v) = \int_{t=-\Delta t/2}^{t=+\Delta t/2} A \, e^{2\pi i v_1 t} e^{-2\pi i v t} \, dt$$

$$= A \int_{t=-\Delta t/2}^{t=+\Delta t/2} e^{-i 2\pi (v - v_1) t} \, dt$$

$$= \frac{A}{-i2\pi(v - v_1)} \left[e^{-i 2\pi (v - v_1) t} \right]_{-\Delta t/2}^{\Delta t/2}$$

whence

$$F(v) = A \, \Delta t \left(\frac{\sin \pi (v - v_1) \Delta t}{\pi (v - v_1) \Delta t} \right) . \tag{G.02}$$

Details of this sinc function are shown in Fig. G.01(c).

Our truncated wave is therefore not monochromatic. Looking at the intercepts on the v-axis we see that only if Δt is infinite can the wave be

$$f(t) = A \cos 2\pi\nu_1 t$$

(a)

(b)

(c)

Fig. G.01

monochromatic. Otherwise, it consists of a spread of frequencies about v_1, mainly in the range $\Delta v = v - v_1 = 1/\Delta t$ (cf. the bandwidth theorem $\Delta v \Delta t \sim 1$ (§4.7)).

It is interesting to note that this result can be arrived at by using the convolution theorem. The continuous cosine wave is the Fourier transform of a single frequency (v_1), and a sinc function is the transform of a rectangle function. The convolution theorem tells us that the product of the cosine wave and the rectangle function, i.e. the truncated cosine wave, is the convolution of the single frequency with the sinc function, and that is what we have in Fig. G.01(c).

APPENDIX H — MAXIMUM ENTROPY DATA PROCESSING

S. F. Gull

The way in which the advent of powerful digital computers has revolutio-
nized the processing of images is mentioned in many places in this book.
When pictures are converted into lists of numbers in a computer, many
complicated operations become feasible that were impossible in direct
image formation. Indeed, computers first became important during the
1950s and 1960s in fields such as radioastronomy where direct imaging was
not even possible; data are collected and converted to digital form and only
then are pictures produced — often, as we have seen, by Fourier transform.
Today's computers can now improve these images further using powerful
new methods of processing. One scheme recently developed for this
purpose, the **maximum entropy method**, is particularly general in its concept
and has come into widespread use in many fields.

The maximum entropy method is based on the statistical mechanics
developed in the last century by Gibbs, but its modern development is
largely the work of Jaynes (1983). Applied to image processing, it represents
a fundamental shift in the way scientists view the problem. Rather than
attempt to manipulate the data directly to form an image, we search through
all possible object distributions, or 'scenes', to find the one that is most likely
to have produced the observed data. Suppose we have a set of data $\{D_k\}$
relevant to a scene that has been divided into picture elements (pixels) and
that the illuminance in these pixels is a set of numbers $\{f_i\}$. If matrix $\mathbf{O}$
represents the connection (assumed linear) between a trial scene and the
(mock) data it would produce, we can write:

$$D_k = \sum_i \mathbf{O}_{ki} f_i + r_k .$$

In some applications this connection would be a Fourier transform (as in
radioastronomy and NMR tomography), or a set of projections (X-ray
tomography), or a convolution with a point-spread function (optical picture
processing). We also allow for the data being corrupted by a noise vector
$\{\mathbf{r}_k\}$. Because the experimentally observed data are in general incomplete
and noisy, there are many possible scenes that would produce mock data in
statistical agreement with the actual data. In practice we usually express this
in terms of a single chi-squared criterion:

$$\sum_k (D_k - \mathbf{O}_{ki} f_i)^2 / \sigma_k^2 \leq (\text{number of data points}) ,$$

where $\{\sigma_k^2\}$ is the variance of the noise. All scenes that satisfy this criterion are allowed by the data, and we must choose between those that remain on other grounds.

The way in which maximum entropy selects a single image may be viewed in terms of a simple model for the production of the underlying scene: we assume that it is generated randomly. Suppose that the total illuminance of the scene is divided into a large number N of small grains and these are distributed amongst the pixels by the traditional 'team of monkeys' to produce random scenes. Most scenes produced in this way are incompatible with the data and these we reject. Those scenes that are compatible are then sorted into piles and the process is (conceptually!) continued until every conceivable, compatible scene has occurred many times over. We then select the largest pile as representing the most likely scene allowed by the data.

The relation to statistical mechanics comes about because our monkeys produce various scenes with different probabilities. If $\{n_i\}$ are the number of grains of illuminance falling into the pixels, then the number of ways in which a given scene can occur is given by the degeneracy $N!/\Pi n_i!$. Taking logarithms and using Stirling's approximation for large N, we see that the largest pile will be that with the greatest entropy:

$$ S = -\sum_i p_i \log p_i \quad \text{where} \quad p_i = n_i/N = f_i \bigg/ \sum f \, . $$

The scene produced most often by the monkeys (the global maximum of entropy) has a uniform intensity and corresponds to a blank picture. We can therefore say that the maximum entropy image is the most uniform one that is permitted by the data. Features will only appear if there is definitive evidence for them in the data, and this property makes the method uniquely safe. Indeed, it can be shown that maximum entropy is the *only* data analysis technique that will not, by itself, introduce spurious correlations into the final picture. It was first applied to radioastronomical image processing by Gull and Daniell (1978) but, with the availability of robust algorithms for the calculation of maximum entropy images, is now a standard tool in many different disciplines. For more details and examples the reader is referred to the review by Gull and Skilling (and references therein).

APPENDIX I — APPLICATIONS IN COMMUNICATIONS
J. D. Weaver

The concepts of frequency domain analysis, convolution and correlation, developed in the main text in relation to optics, are — as mentioned in several places — fundamental also in Communications. In a Communications channel, a signal $s(t)$ (a function of time) is received by an observer, having passed through an intervening 'system' comprising the transmission path and, in general, appropriate transmitting and receiving equipment. The observer's problem is to recover the original signal from the version, $r(t)$, that he receives.

Consideration of the problem in the frequency domain leads to an apparently simple solution. The transmitted signal has a Fourier spectrum $S(f)$ (cf. Eqn (4.33)), and the received signal a Fourier spectrum $R(f)$. The intervening system is characterized by a **frequency response**, $H(f)$, a function giving the change in amplitude and phase of a single input sinusoid of frequency f; it is equivalent to the **transfer function** of §5.1. This function is the Fourier transform of the **impulse response** of the system, which is equivalent in this application, to the **line spread function** in optics (§§4.4, 4.5).

The convolution theorem (Eqn (4.25)) relates the three functions: $R(f) = S(f) H(f)$. This can be inverted to give $S(f)$ if $H(f)$ is known:

$$S(f) = R(f)/H(f) . \qquad (I.01)$$

$s(t)$ can then be recovered by Fourier transformation of $S(f)$.

Practical problems arise which complicate this technique. For all real systems, $H(f)$ has ranges of frequency for which it is zero or very small. $R(f)$ will be small in the same ranges, and the ratio is either indeterminate or subject to large errors as a result of observational errors in both $R(f)$ and $H(f)$.

This particular problem is overcome by insisiting that $H(f)$ shall not become small except at frequencies for which $S(f)$ is negligible. It is then unnecessary to evaluate $S(f)$ at frequencies where $H(f)$ is small. This requirement is normally expressed by saying that the **bandwidth** of the transmission system must exceed the signal bandwidth (the bandwidth of a function of time being the range of frequencies for which its Fourier transform has significant values).

A more serious complication in the use of Eqn (I.01) is that the received signal always contains a component — **noise** — which originates not in the transmitted signal, but in the transmission system. There are numerous sources of noise (lightning discharges ('atmospherics'), ionospheric disturbances, electronic noise due to thermal motion of electrons, etc.) which have one characteristic in common — they are random, unpredictable, and can be specified only statistically.

The received spectrum therefore comprises a component $S(f)\,H(f)$ due to the signal, and a noise component $N(f)$:

$$R(f) = S(f)\,H(f) + N(f)\ ,$$

so that

$$S(f) = \frac{R(f)}{H(f)} - \frac{N(f)}{H(f)}\ . \qquad (\text{I.02})$$

Since $N(f)$ is known at best statistically, there is no straightforward means of evaluating the second term in this equation. Only if $N(f)$ is small compared with $S(f)$ throughout the signal bandwidth can a good estimate of $S(f)$ be obtained.

For cases where $N(f)$ is not small, special techniques have been developed for recovering $S(f)$ and $s(t)$ with minimum error. These techniques, which involve convolution and correlation, do, however, require some prior knowledge of the signal. A brief look at two such techniques will give an understanding of some of the ideas involved.

Suppose that a signal $s(t)$, if it is being transmitted, is periodic with period T. Suppose also that the noise in the system is so large that the preceding technique does not adequately recover the signal. Let the received signal plus noise be represented by $\rho(t)$, so $\rho(t) = r(t) + n(t)$, where $n(t)$ is the noise, and $r(t)$ originates from the transmitted signal. Suppose $\rho(t)$ is measured at time τ and at multiples of T thereafter, N measurements being taken. The set of measurements taken are

$$\rho(\tau),\ \rho(\tau + T),\ \rho(\tau + 2T),\ \ldots,\ \rho(\tau + (N-1)T)$$

Let these be averaged:

$$\overline{\rho(\tau)} = \frac{1}{N} \sum_{k=0}^{N-1} \rho(\tau + kT)$$

$$= \frac{1}{N} \left[\sum_{k=0}^{N-1} r(\tau + kT) + \sum_{k=0}^{N-1} n(\tau + kT) \right]\ .$$

Now if $s(t)$ is periodic, so too will be $r(t)$, and

$$\sum_{k=0}^{N-1} r(\tau + kT) = Nr(\tau)\ ;$$

therefore

$$\overline{\rho(\tau)} = r(\tau) + \frac{1}{N}\sum_{k=0}^{N-1} n(\tau + kT) \ .$$

Now $n(t)$ is a random signal, and the values of n in the summation will not all be the same; some will be positive, some negative, and their sum will grow at a rate not proportional to N, but more like $N^{\frac{1}{2}}$, depending on the details of the noise statistics. The noise term is therefore proportional to $N^{-\frac{1}{2}}$ and becomes smaller as more samples are included. So for large N, we have $\overline{\rho(\tau)} = r(\tau)$. If this process is repeated for all τ in the range 0 to T, the complete waveform of $r(t)$ is obtained. $s(t)$ can be recovered if necessary by the Fourier method described above.

This technique for recovering $r(t)$ can be carried out in practice by multiplying the received waveform $\rho(t)$ by a sampling waveform $\Delta(t-\tau)$ which in principle consists of unit impulse functions (§4.3) at τ and multiples of T. The impulse functions select the values of received signal-plus-noise at the instants of their occurence (i.e. at $\tau + kT$, as required) and subsequent averaging should yield $r(\tau)$. That this is so can easily be established, yielding Eqn (I.04).

The sampling waveform is given by

$$\Delta(t-\tau) = \sum_{k=0}^{N} \delta(t - \tau - kT)$$

(remember that the impulse function $\delta(t-t')$ occurs at $t = t'$). Multiplying by $\rho(t)$ and averaging the product, by integration and division by time, gives

$$\overline{\rho\Delta} = \frac{1}{NT}\int_{\tau}^{\tau+NT} \Delta(t-\tau)\rho(t)\,\mathrm{d}t \qquad (I.03)$$

$$= \frac{1}{NT}\int_{\tau}^{\tau+NT} \rho(t)\sum_{k=0}^{N}\delta(t-\tau-kT)\,\mathrm{d}t \ .$$

Interchange the order of summation and integration:

$$\overline{\rho\Delta} = \frac{1}{NT}\sum_{k=0}^{N}\int_{\tau}^{\tau+NT} \rho(t)\delta(t-\tau-kT)\,\mathrm{d}t \ .$$

Now $\delta(t-\tau-kT)$ is non-zero only at $t = \tau + kT$. We can therefore replace $\rho(t)$ by $\rho(\tau+kT)$ and remove it from the integral:

$$\overline{\rho\Delta} = \frac{1}{NT}\sum_{k=0}^{N}\rho(\tau+kT)\int_{\tau}^{\tau+NT} \delta(t-\tau-kT)\,\mathrm{d}t \ .$$

By definition the integral of an impulse function is unity (§4.3) and (I.04) is obtained.

$$\overline{\rho\Delta} = \frac{1}{NT} \sum_{k=0}^{N} \rho(\tau + kT)$$

$$= \frac{N+1}{NT} r(\tau) + \frac{1}{NT} \sum_{k=0}^{N} n(\tau + kT) \ . \qquad (I.04)$$

The first term is, as required, proportional to the received signal $r(\tau)$, while the second (noise) term reduces as N becomes larger. Multiplication by the sampling waveform and averaging does therefore produce the value of the received signal with reduced noise, essentially by making use of partial self-cancellation of noise. Notice, in particular, the fundamental equation (I.03) — this is the **cross-correlation function** (§4.7) of $\rho(t)$ and $\Delta(t)$, for the delay τ.

A basic radar system is a good example of the cross-correlation technique just described. A periodic train of pulses of radio frequency is transmitted. Some of the energy is reflected by objects within range, and received on the same aerial. The sampling waveform is derived from the transmitted pulses, with suitable variable delay τ inserted electronically. Cross-correlation with the incoming signal reveals both the presence of reflections and their time of reception after the transmission of a pulse.

If the signal is not periodic, it is still possible to discover its presence in noise, provided that the shape of the signal to be received, $r(t)$, is known, at least approximately, beforehand. Suppose that $r(t)$, if present, has duration T. Let a number of samples of $\rho(t)$, signal-plus-noise, be taken at intervals δt in period T. The average of these samples is

$$\overline{\rho} = \frac{1}{N+1} \sum_{k=0}^{N} \rho(k\delta t)$$

$$= \frac{1}{N+1} \sum_{k=0}^{N} r(k\delta t) + \frac{1}{N+1} \sum_{k=0}^{N} n(k\delta t) \ .$$

In this case, the values of $r(k\delta t)$ are not all the same, and they can be of opposite sign, so the signal and noise terms do not necessarily build up at markedly different rates. To overcome this difficulty, the sample values are multiplied by the previously known expected values of r. Since it is not known how the beginning of the sampling period relates in time to the beginning of the signal waveform, a delay time is incorporated. The average of $\rho(t)r(t-\tau)$ is formed:

$$\overline{\rho r} = \frac{1}{N+1} \sum_{k=0}^{N} \rho(k\delta t) \ r(k\delta t - \tau)$$

$$= \frac{1}{N+1} \sum_{k=0}^{N} r(k\delta t)r(k\delta t - \tau) + \frac{1}{N+1} \sum_{k=0}^{N} n(k\delta t)r(k\delta t - \tau) \ .$$

The second term is roughly proportional to $N^{-\frac{1}{2}}$, because there is no relationship between $n(t)$ and $r(t)$ and the individual products in the summation are random, positive or negative. The first term is expected to build up rapidly for $\tau = 0$, when it reduces to

$$\frac{1}{N+1} \sum_{k=0}^{N} r^2(k\delta t) \ ,$$

because all the terms are positive. But for other values of τ the build-up will be less rapid because the product can then take on negative as well as positive values. It is found that $\overline{\rho r}$ is small for large τ, but increases rapidly as τ approaches 0. A large value of the average at a particular value of τ is taken as an indication of the presence of a signal at this value of time.

The average takes on a familiar form if $\delta t \to 0$. Then $k\delta t$ becomes a continuous function t, and $N + 1 \to N = T/(\delta t) \to T/(\mathrm{d}t)$. So

$$\overline{\rho r} \to \frac{1}{T} \int_{0}^{T} \rho(t)r(t-\tau)\,\mathrm{d}t \ . \tag{I.05}$$

Remember that r in this equation is the known received waveform. The multiplication is performed by passing $\rho(t)$ through a filter whose impulse response contains $r(t)$.

Now if a signal $\rho(t)$ is passed through a filter with impulse response $h(t)$, the output $o(t)$ is obtained by convolution as

$$o(t) = \int \rho(t)h(\tau - t)\,\mathrm{d}t \ . \qquad\qquad \text{cf. Eqn (4.25)}$$

Comparing this with (I. 05) shows that the integral there is the **convolution** of the received signal-plus-noise with a filter whose impulse response is the *time-reversed* signal waveform. Such a filter is called a **matched filter,** because its impulse response is tailored to suit a particular signal, and it is the most effective filter for detecting that particular signal.

An important comment must be made concerning the limiting process $\delta t \to 0$. In this process, the number of samples of signal-plus-noise becomes infinite, and it would appear that the noise term should become zero. This does not happen in practice for the following reason. For the noise samples to give mutual cancellation they must be statistically independent — the value of one sample must have no influence on the values to be observed for the others. It can be shown that this requires the samples to be taken at intervals $\delta t > 1/2f_{max}$, where f_{max} is the maximum frequency passed by the system. For $\delta t \to 0$ an infinite bandwidth would be required for all samples to be independent, which is impossible in practice. In proceeding to the limit, non-independent samples are being included, and the noise cancellation is less than perfect.

One question has been left unanswered. How big does the convolution integral have to be for it to be decided that a signal is present? The answer is statistical, for no matter how big the value of the convolution, it is always possible that it occurred due to noise alone; but it is less likely the bigger the convolution integral. It is not too difficult to derive a *decision level*, which is such that, if the filter output exceeds it, then a signal is present with a predetermined probability.

The illustrative examples outlined above demonstrate the relevance of the material in the main text, particularly in Chapter 4, to fields other than optics. The signal processing involved has been described here in terms of continuous, or analogue signals. Increasingly, however, signal processing is being carried out by computers operating on sampled versions of signals. The principles remain unchanged, but the versatility and power of the digital computer do permit the use of much more advanced techniques than those developed here (see, for example, Appendix H).

Further reading

Lynn, P. A. (1973) *An introduction to the analysis and processing of signals.* Macmillan, London.

Schwartz, M. and Shaw, L. (1975) *Signal processing.* McGraw-Hill, New York.

Weaver, J. D. (in preparation) *Fundamentals of signal processing.*

APPENDIX J — THE IMPULSE RESPONSE OF THE EYE:
A DIFFRACTION ANALOGY
Janet E. Wolf

It is fascinating to see patterns normally associated with Fraunhofer diffraction appearing in analogous but otherwise totally different situations. A rotating striped pattern of parallel lines (Fig. J.01) produces the odd visual

Fig. J.01 — A pattern consisting of a set of parallel stripes with the origin of the Cartesian coordinate system at the centre of rotation and the x, y axes fixed on the pattern. The pattern is rotated with a constant angular velocity ω, and the point $P(x, y)$ travels with a constant velocity ωr. In the cutaway, we see that the component of this velocity perpendicular to the stripes is $\omega r \cos \theta = \omega x$. It is this component that produces the blurring of the stripes (see text).

effect of a clearly resolved central band of 'heightened contrast' nearly perpendicular to the stripes and rotating with the pattern as a whole. Still photographs of the rotating pattern display similar effects (Fig. J.02). It can be shown that the contrast across the perceived band (or the recorded contrast in the case of photographs) bears a Fourier transform relation to the temporal impulse response in the case of the eye, or the 'exposure function' in the case of the camera. We find that the impulse response is related to the perceived contrast in the same way that the aperture function is related to the amplitude distribution in a Fraunhofer diffraction pattern.

In photographs of scenes containing moving objects, a car say, we are used to the image of the moving object being blurred unless a fast shutter speed is used. This is easily understood if we think of the film as summing or integrating a succession of images displaced in time and space for as long as the aperture is open — a faster movement produces a greater smear. An

Fig. J.02 — A single photographic exposure of a rotating set of parallel stripes such as those shown in Fig. J.01. The perceived band has a contrast distribution across it that is in the form of a sinc function when the exposure function is a rectangle function.

analogous smearing effect can also occur with the human visual system even though the processes are different. Unlike a camera, when we are observing a scene we keep our eyes open. If the scene contains a moving object and we do not follow it with our eyes the image of the object travels across the retina leaving a trail of persisting images. This persistence is part of the perceptual process and the exact manner in which the momentary retinal image builds up and decays is called the **temporal impulse response** of the eye. The perceived spatial smear is a manifestation of the temporal response of the eye, and observation of the former can reveal information about the latter.

Consider a set of parallel stripes rotating with constant angular velocity, ω (Fig. J.02). Only the component of velocity perpendicular to the stripes will cause their blurring and it is easily seen that this velocity is constant (ωx) for any distance x from the centre, independent of y. This means that we have a central parallel-sided band which is increasingly smeared away from the centre, the ultimate boundary being determined by the angular velocity and the impulse response of the eye (or camera exposure function).

Contrast distribution across the band (along the x-axis) for an idealized rectangular impulse response of duration T

Consider the case of a luminance distribution across the stationary stripes that is sinusoidal and given by

$$I(y) = \alpha + \beta \cos(2\pi y/2d) . \tag{J.01}$$

At a given distance x from the centre of rotation the sinusoidal profile advances perpendicularly to the stripes with a velocity ωx, causing smearing over a distance $\omega x T$ (Fig. J.03(i)). The perceived luminance in the band at

Fig. J.03 — Illustration of the underlying analogy between the perceived contrast of the rotating stripes as a function of the distance x from the y-axis, and the amplitude distribution in a Fraunhofer diffraction pattern as a function of θ.
  (i) The smearing by the rectangle function $\omega x T$ reduces the contrast of the stripes.
  (ii) The resultant amplitude of the light diffracted in a direction θ for a rectangular aperture function is reduced as a consequence of displacement (analogous to the smearing in (i)) of wavelets leaving successive points across the aperture.

that value of x is given by

$$I'(y) = \int_{-\omega x T/2}^{+\omega x T/2} I(y)\,dy \ . \tag{J.02}$$

This is reminiscent of the smearing of fringes in interferometry due to finite source width (cf. Fig. 7.01). It leads to a perceived contrast, C, given (cf. Eqn (7.11)) by

$$C = \frac{I'_{\max} - I'_{\min}}{I'_{\max} + I'_{\min}} = \frac{\beta}{\alpha}\,\frac{\sin(\pi\omega x T/2d)}{(\pi\omega x T/2d)} \ . \tag{J.03}$$

Since the contrast of the stationary pattern, C_0, is given via Eqn (J.01) by

$$C_0 = \frac{I_{\max} - I_{\min}}{I_{\max} + I_{\min}} = \frac{\beta}{\alpha} \tag{J.04}$$

we have

$$C = C_0 \frac{\sin(\pi\omega x T/2d)}{(\pi\omega x T/2d)} \ . \tag{J.05}$$

This contrast distribution is nicely illustrated in Fig. J.01 where we clearly see it following the sinc function. As expected for a sinc function, the width of the main band is twice that of the secondary bands, the phase of the second band being the reverse of the first (i.e. a black stripe is opposite a white stripe), the third is the same, and so on.

For this idealized rectangular impulse response we therefore have the result that the contrast distribution across the perceived band is the Fourier transform of the impulse response. Similarly for a more general form of impulse response the contrast across the perceived band would again bear the transform relation to the temporal impulse response.

The diffraction analogy

The analogy with Fraunhofer diffraction is now apparent. The amplitude distribution across the diffraction pattern due to a single slit of rectangular aperture function, width a, is given by

$$R_1 = R_1(0)\left[\frac{\sin \pi u a}{\pi u a}\right] \qquad \{2.05\} \; (\text{J}.06)$$

and is the Fourier transform of the aperture function.

The analogy can be appreciated qualitatively by reference to Fig. J.03. In examining the analogy we see that the aperture function is analogous to the impulse response function. In particular, for the rectangle function, 'a' is equivalent to T. Also, the spatial frequencies $u = (\sin \theta)/\lambda$ that constitute the spectrum of the aperture function observed as a function of θ have been replaced by the temporal frequencies, $v = \omega x/2d$, that constitute the spectrum of the impulse response and are represented or provided by the target at different values of x. There has in effect been a complete exchange of spatial and temporal factors. Finally, the phase associated with each spatial frequency in the diffraction pattern has its equivalence in the band effect. Any delay of a particular temporal frequency or group of frequencies results in a rotation of that localized section of the band. This would show as a kink in the straight lines within the band. The magnitude of the kink at any value of x, associated with a temporal frequency $\omega x/2d$, is proportional to the phase shift for that frequency. Hence the shape of the lines represents the phase transfer frunction of the system.

The diffraction analogy can now be used to explain the striking visual effects seen when rotating stripes are viewed under stroboscopic illumination (Fig. J.04). The central band seen without strobing is still there but additional, identical bands are seen at equal intervals across the pattern. The explanation is simply that the flashing light is sampling the impulse response in the same way as the slits in a diffraction grating can be thought of as sampling the overall aperture function. Once again the contrast represents the Fourier transform of the sampled impulse response.

Fig. J.04 — Photographs of the rotating stripes stroboscopically illuminated. (i)–(iii), taken with a fixed exposure time but with the flash interval successively decreasing (indicated by the marks above the pictures). (iv)–(vi), taken with a fixed flash interval but with the exposure times successively increasing. (The increasing number of flashes within the exposure time is indicated by the marks above the pictures.)

In recognizing that we can think of the stroboscopic illumination as sampling the temporal impulse response, rather than the more conventional approach of thinking of it as sampling the moving target, this repeating band effect is providing us with a powerful tool for studying the phenomenon of 'apparent movement', namely the appearance of movement that results from a series of discrete successively displayed images as found in cinema-photography or VDU displays of moving targets.

However, the most significant feature associated with this band effect (unstrobed) is that we see displayed not only the temporal modulation transfer function but also the phase transfer function. This means the effect is providing us for the first time with a method of establishing the complex temporal transfer function, thereby enabling us to establish the temporal impulse response of the eye.

APPENDIX K — THE ELECTROMAGNETIC SPECTRUM

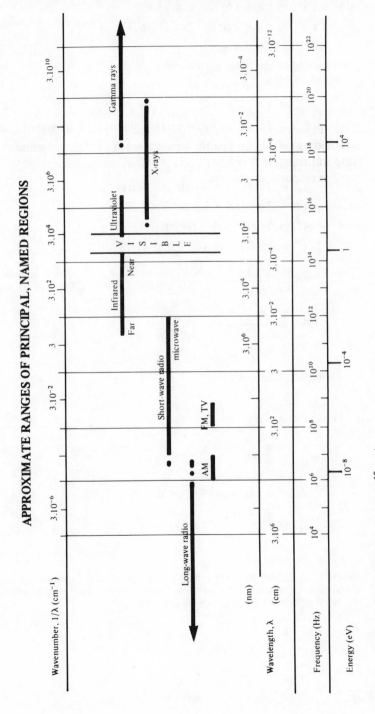

APPROXIMATE RANGES OF PRINCIPAL, NAMED REGIONS

Velocity taken as approximately 3.10^{10} cm/s.
1 eV has associated frequency c. 2.42×10^{14} Hz,   wavelength c. 1240 nm.

APPENDIX L — USEFUL FORMULAE

$$\sin(A \pm B) = \sin A \, \cos B \pm \cos A \sin B$$

$$\cos(A \pm B) = \cos A \, \cos B \mp \sin A \sin B$$

$$\sin 2\theta = 2 \sin \theta \cos \theta$$

$$\cos 2\theta = \cos^2 \theta - \sin^2 \theta$$

$$= 2\cos^2 \theta - 1$$

$$= 1 - 2\sin^2 \theta$$

$$\sin(N+1)\theta = 2\cos\theta \, \sin N\theta - \sin(N-1)\theta$$

$$\cos(N+1)\theta = 2\cos\theta \, \cos N\theta - \cos(N-1)\theta$$

Chebyshev recursion formulae

$$\sin A + \sin B = 2\sin\tfrac{1}{2}(A+B) \, \cos\tfrac{1}{2}(A-B)$$

$$\sin A - \sin B = 2\cos\tfrac{1}{2}(A+B) \, \sin\tfrac{1}{2}(A-B)$$

$$\cos A + \cos B = 2\cos\tfrac{1}{2}(A+B) \, \cos\tfrac{1}{2}(A-B)$$

$$\cos A - \cos B = -2\sin\tfrac{1}{2}(A+B) \, \sin\tfrac{1}{2}(A-B)$$

$$e^{i\theta} = \cos\theta + i\sin\theta \quad \text{(Euler's relation)}^\dagger$$

$$\cos\theta = \frac{e^{i\theta} + e^{-i\theta}}{2}$$

$$\sin\theta = \frac{e^{i\theta} - e^{-i\theta}}{2i}$$

$$\int \sin x \, dx = -\cos x$$

$$\int \cos x \, dx = \sin x$$

$$\int e^{ax} \, dx = \frac{1}{a}e^{ax}$$

$$S_n = a + ar + ar^2 + \ldots + ar^{n-1}$$

$$= \frac{a(1-r^n)}{1-r} \quad (r \neq 1)$$

---

† Described by Richard Feynman as 'this amazing jewel ... the most remarkable formula in mathematics'. (Feynman was joint Nobel Prizewinner in physics in 1965 for his contribution to the development of quantum electrodynamics.)

References

Abbe, E. (1873) Beiträge zur theorie des mikroskops und der mikroskopis-
chen wahrnehmung. *Archiv. f. Microskop. Anat.*, **9**, 413–468. (*Collected Works* (1904) Vol. 1.)

Barer, R. (1955) Phase contrast microscopy. *Research*, **8**, 341–343.

Blythe, J. H. (1957) A new type of pencil beam aerial for radio astronomy.
Roy. Ast. Soc. Monthly Notices, **117**, 644–651.

Bolton, J. G. and Stanley, G. J. (1948) Variable source of radio frequency
radiation in the constellation of Cygnus. *Nature*, **161**, 312–313.

Bracewell, R. N. (1956) Strip integration in radio astonomy. *Aust. J. Phys.*,
9, 198–217.

Bragg, W. L. (1929) The determination of parameters in crystal structures
by means of Fourier analysis. *Proc. Roy. Soc. A*, **123**, 537–559.

Bragg, W. L. (1929) An optical method of representing the results of X-ray
analysis. *Z. f. Krist.* **70**, 475–492.

Bragg, W. L. (1933) *The crystalline state*, Vol. I. Bell, London.

Bragg, W. L. (1939) A new type of 'X-ray microscope'. *Nature*, **143**, 678.

Bragg, W. L. (1942) The X-ray microscope. *Nature*, **149**, 470–471.

Bragg, W. L. (1944) Lightning calculations with light. *Nature*, **154**, 69–72.

Brooks, R. A. and Di Chiro, G. (1975) Theory of image reconstruction in
computed tomography. *Radiology*, **117**, 561–572.

Brooks, R. A. and Di Chiro, G. (1976) Principles of computer assisted
tomography (CAT) in radiographic and radioisotopic imaging. *Phys.
Med. Biol.*, **21**, 689–732.

Buerger, M. J. (1950) Generalised microscopy and the two-wavelength
microscope. *J. Appl. Phys.*, **21**, 909–917.

Cannon, T. M. (1976) Blind deconvolution of spatially invariant image blurs
with phase. *IEEE Trans ASSP*, **24**.

Carson, P. L., Oughton, T. V., Hendee, W. R., and Ahuja, A. S. (1977)
Imaging soft tissue through bone with ultrasound transmission tomography by reconstruction. *Med. Phys.*, **4**, 302–309.

Carson, P. L. and Scherzinger, A. L. (1980) Ultrasonic computed tomography. In Wells, P. N. T. and Zoskin, M. C. (eds), *New techniques and instrumentation in ultrasonography.* Churchill Livingstone, New York.

Connes, P. (1971) Advances in Fourier spectroscopy. *Phys. Bull.*, **22**, 26–28.

Cooley, J. W. and Tukey, J. W. (1965) An algorithm for the machine calculation of complex Fourier series. *Mathematics of Computing*, **19**, 297–301.

Cormack, A. M. (1963) Representation of a function by its line integrals, with some radiological applications. *J. Appl. Phys.*, **34**, 2722–2727.

DeRosier, D. J. and Klug, A. (1968) Reconstruction of three-dimensional structures from electron micrographs. *Nature*, **217**, 130–134.

DeRosier, D. J. and Klug, A. (1972) Structure of the tubular variants of the head of bacteriophage T4 (polyheads). *J. Mol. Biol.*, **65**, 469–488.

Duck, F. A. and Hill, C. R. (1979) Mapping true ultrasonic backscatter and attenuation distributions in tissue — a digital reconstruction approach. In Linzer, M. (ed.), *Ultrasonic tissue characterization II, NBS* Spec. Pub. 525 Washington DC, US Govt. Ptg. Office, 247–251.

Elias, P. (1953) Optics and communication theory. *Jour. Opt. Soc. Amer.*, **43**, 229–232.

Elias, P., Grey, D. S. and Robinson, D. Z. (1952) Fourier treatment of optical processes. *Jour. Opt. Soc. Amer.*, **42**, 127–134.

Fellgett, P. (1951) PhD thesis, University of Cambridge, UK.

Fellgett, P. (1958) A propos de la théorie du spectromètre interférentiel multiplex. *J. Phys. Radium*, **19**, 187.

Fizeau, H. (1862) *Ann. Chim. Phys.*, **66**, 429.

Fizeau, H. (1868) *C. R. Acad. Sci. Paris*, **66**, 934.

Gabor, D. (1948) A new microscope principle. *nature*, **161**, 777–778.

Gordon, R., Bender, R., and Herman, G. T. (1970) Algebraic reconstruction techniques (ART) for three-dimensional electron microscopy and X-ray photography. *J. Theoret. Biol.*, **29**, 471–481.

Greenleaf, J. F. (1983) Computerized tomography with ultrasound. *Proc. IEEE*, **73(3)**, 330–337.

Greenleaf, J. F., Johnson, S. A., Lee, S. L., and Wood, E. H. (1974) Algebraic reconstruction of spatial distributions of acoustic absorption within tissue from their two-dimensional acoustic projections. In *Acoustic holography*, **5**, 591–603, Plenum Press, New York.

Greenleaf, J. F., Johnson, S. A., Samayoa, W. F., and Duck, F. A. (1975) Algebraic reconstruction of spatial distributions of acoustic velocities in tissue from their time-of-flight profiles. In *Acoustic holography*, **6**, 71–90, Plenum Press, New York.

Gull, S. F. and Daniell, G. J. (1978) Image reconstruction from incomplete and noisy data. *Nature*, **272**, 686–690.

Gull, S. F. and Skilling, J. (1985) Maximum entropy method in image processing. *IEE Proc.*, **131**, 646–659.

Hanbury Brown, R. and Jennison, R. C. (1950) Unpublished.

Hanbury Brown, R. and Twiss, R. Q. (1954) A new type of interferometer for use in radio astronomy. *Phil. Mag.*, **45**, 663–682.

Hanbury Brown, R. and Twiss, R. Q. (1956) A test of a new type of stellar interferometer on Sirius. *Nature*, **178**, 1046–1048.

Hanbury Brown, R. and Twiss, R. Q. (1958) Interferometry of the intensity fluctuations in light — III. Applications to astronomy. *Proc. Roy. Soc. A*, **248**, 199–221.

Hanson, A. W. and Lipson, H. (1952) A simplified fly's eye procedure. *Acta Cryst.*, **5**, 145.

Harburn, G., Taylor, C. A., and Welberry, T. R. (1975) *Atlas of optical transforms*. Bell, London.

Hopkins, H. H. (1951) The concept of partial coherence in optics. *Proc. Roy. Soc. A*, **208**, 263–277.

Hounsfield, G. N. (1973) Computerised transverse axial scanning (tomography) Part 1. Description of system. *Brit. J. Radiol.*, **46**, 1016–1022.

Jacquinot, P. (1960) New developments in interference spectroscopy. *Rep. Progr. Phys.*, **23**, 267–312.

Jacquinot, P. and Dufour, C. J. (1948) Conditions optiques d'emploi des cellules photo-électriques dans les spectrographes et les interféromètres. *J. Rech. C.N.R.S.*, **6**, 91–103.

Janes, E. T. (1983) *Papers on probability, statistics ana statistical physics*. Synthese Library, Vol. 158 (ed. R. D. Rosenkrantz) Reidel, Dordrecht.

Jennison, R. C. (1958) A phase sensitive interferometer technique for the measurement of the Fourier transforms of spatial brightness distributions of small angular extents. *MNRAS*, **118**, 276–284.

Kak, A. C. (1979) Computerised tomography with X-ray, emission, and ultrasound sources. *Proc. IEEE*, **67**, 1245–1272.

Klug, A. (1978–79) Image analysis and reconstruction in the electron microscopy of biological macromolecules. *Chemica Scripta*, **14**, 245–256.

Klug, A. and Berger, J. E. (1964) An optical method for the analysis of periodicities in electron micrographs, and some observations on the mechanism of negative staining. *J. Mol. Biol.*, **10**, 565–569.

Klug, A. and Crowther, R. A. (1972) Three-dimensional image reconstruction from the viewpoint of information theory. *Nature*, **238**, 435–440.

Klug, A. and Finch, J. T. (1968) Structure of viruses of the papilloma-polyoma type. *J. Mol. Biol.*, **31**, 1–12.

Kuhl, D. E. and Edwards, R. Q. (1963) Image separation radioisotopic scanning. *Radiology*, **80**, 653–661.

Kuhl, D. E., Hale, J. and Eaton, W. L. (1966) Transmission scanning: a useful adjunct to conventional emission scanning for accurately keying isotope deposition to radiographic anatomy. *Radiology*, **87**, 278–284.

Lauterbur, P. C. (1973) Image formation by induced local interactions: examples employing nuclear magnetic resonance. *Nature*, **242**, 190–191.

Ledley, R. S., Di Chiro, G., Luessenhop, A. J., and Twigg, H. L. (1974) Computerized transaxial X-ray tomography of the human body. *Science, NY*, **186**, 207–212.

Lehar, A. F., Preston, F. H., and Cannon, T. M. (1982) An image processing system for enhancement and deblurring of photographs. *Proceedings of IEE conference on electronic image processing.* York, UK. pp. 174–178.

Leith, E. N. and Upatnieks, J. (1962) Reconstructed wavefronts and communication theory. *J. Opt. Soc. Amer.*, **52**, 1123–1130.

Leith, E. N. and Upatnieks, J. (1963) Wavefront reconstruction with continuous-tone objects. *J. Opt. Soc. Amer.*, **53**, 1377–1381.

Leith, E. N. and Upatnieks, J. (1964) Wavefront reconstruction with diffused illumination and 3-dimensional objects. *J. Opt. Soc. Amer.*, **54**, 1295–1301.

Leith, E. N. and Upatnieks, J. (1965) Wavefront reconstruction photography. *Physics Today*, **18**, No. 8, 26–32.

Michelson, A. A. (1890a) On the application of interference methods to astronomical measurements. *Phil. Mag.*, Ser 5, **30**, 1–21.

Michelson, A. A. (1890b) Measurement by light-waves. *Am Jnl. Sci.*, **39**, 115–121.

Michelson, A. A. (1891a) Visibility of interference-fringes in the focus of a telescope. *Phil. Mag.*, Ser. 5, **31**, 256–259.

Michelson, A. A. (1891b) On the application of interference-methods to spectroscopic measurements — I. *Phil. Mag.*, Ser 5, **31**, 338–346.

Michelson, A. A. (1892) On the application of interference-methods to spectroscopic measurements — II. *Phil. Mag.*, Ser 5, **34**, 280–299.

Michelson, A. A. (1920) On the application of interference methods to astronomical measurements. *Astrophys. Jnl*, **51**, 257–262.

Michelson, A. A. (1927) *Studies in optics.* University of Chicago Press, Chicago. (Phoenix Science Series edn. 1962).

Mills, B. Y. and Little, A. G. (1953) A high resolution aerial system of a new type. *Australian J. Phys.*, **6**, 272–278.

Oldendorf, W. H. (1961) Isolated flying spot detection of radiodensity discontinuities — Displaying the internal structural pattern of a complex object. *IEEE Trans. Bio-Med. Electronics*, **BMW-8**, 68–72.

Padawer, J. (1968) The Nomarksi interference-contrast microscope. An experimental basis for image interpretation. *J. Roy. Mic. Soc.*, **88**, 305–349.

Pawsey, J. L., Payne-Scott, R. and McCready, L. L. (1946) Radio-frequency energy from the Sun. *Nature*, **157**, 158–159.

Porter, A. B. (1906) On the diffraction theory of microscopic vision. *Phil. Mag.*, Ser 6, **11**, 154–166.

Pratt, W. K. (1978) *Digital image processing.* Wiley, New York.

Pykett, I. L. (1982) NMR imaging in medicine. *Sci. Amer.*, **246**, 54–64.

Radon, J. (1917) Über die bestimmung von Funktionen durch ihre integralwerte längs gewisser Mannigfaltigeiten. *Ber. Verh. Sächs. Akad. Wiss. Leipzig Math. Phys. Kl.*, **69**, 262–277. (Translation by Analog Corp., 1976: On the determination of functions by their integral values along certain manifolds.)

Ramachandran, G. N. and Lakshminarayanan, A. V. (1971) Three-dimensional reconstruction from radiographs and electron micrographs: application of convolution instead of Fourier transforms. *Proc. Nat. Acad. Sci. USA*, **68**, 2236–2240.

Rayleigh, Lord (1874) On the manufacture and theory of diffraction-gratings. *Phil. Mag.*, **47**, 81–93, 193–205. (*Scientific Papers*, Vol. I, 199–221. CUP, Cambridge.)

Rayleigh, Lord (1892) On the interference bands of approximately homogeneous light: in a letter to Prof. A. Michelson. *Phil. Mag.*, Ser 5, **34**, 407–411. (*Scientific Papers*, Vol. IV, 15–18. CUP, Cambridge.)

Rayleigh, Lord (1896) On the theory of optical images, with special reference to the microscope. *Phil. Mag.*, **42**, 167–195. (*Scientific Papers*, Vol. IV, 235–260. CUP, Cambridge.)

Rayleigh, Lord (1903) On the theory of optical images, with special reference to the microscope. *Journ. R. Micr. Soc.*, 474–482. (*Scientific Papers*, Vol. V, 118–125. CUP, Cambridge.)

Rowley, P. D. (1969) Quantitative interpretation of three-dimensional weakly refractive phase objects using holographic interferometry. *J. Opt. Soc. Am.*, **59**, 1496–1498.

Ryle, M. (1952) A new radio interferometer and its application to the observation of weak radio stars. *Proc. Roy. Soc. A*, **211**, 351–375.

Ryle, M. and Vonberg, D. D. (1946) Solar radiation on 175 Mc/s. *Nature*, **158**, 339–340.

Ryle, M. and Smith, F. G. (1948) A new intense source of radio-frequency radiation in the constellation of Cassiopeia. *Nature*, **162**, 462–463.

Ryle, M. and Hewish, A. (1960) The synthesis of large radio telescopes. *Roy. Ast. Soc. Monthly Notices*, **120**, 220–230.

Schade, O. H. (1948) Electro-optical characteristics of television systems. *RCA Review*, **9**, 5–37, 245–286, 490–530, 653–686.

Shannon, C. (1949) *The mathematical theory of communication*. University of Illinois, Urbana.

Skilling, J. (1981) Algorithms and applications. *Proceedings of Laramie conference on maximum entropy*. University of Wyoming, USA.

Stanier, H. M. (1950) Distribution of radiation from the undisturbed Sun at a wave-length of 60 cm. *Nature*, **165**, 354–355.

Taylor, C. A. and Lipson, H. (1964) *Optical transforms*. Bell, London.

Vander Lugt, A. B. (1963) Signal detection by complex spatial filtering. *Radar Lab., Report* No. 4594-22-T, University of Michigan, Ann Arbor.

Vander Lugt, A. B. (1964) Signal detection by complex spatial fitering. *IEEE Trans. Inform. Theory*, **IT-10**:2.

Wiener, N. (1949) *The extrapolation, interpolation and smoothing of stationary time series*. Technology Press and Wiley, New York.

Woodward, P. M. (1953) *Probability and information theory, with applications to radar*. Pergamon, London. (Reprinted by Artec., Mass., 1980.)

Zernike, F. (1938) The concept of degree of coherence and its applications to optical problems. *Physica*, **5**, 785–795.

Zernike, F. (1942) Phase contrast, a new method for the microscopic observation of transparent objects. *Physica*, **9**, 686–698, 974–986.

Ziskin, M. C. and Wells, P. N. T. (1980) Tissue characterization. In Wells, P. N. T. and Ziskin, M. C. (eds), *New techniques and instrumentation in ultrasonography*. Churchill Livingstone, New York.

Bibliography

GENERAL

Born, M. and Wolf, E. (1980) *Principles of optics*, 6th edn. Pergamon, Oxford.

Brown, E. B. (1965) *Modern optics*. Reinhold, New York.

Ditchburn, R. W. (1976) *Light*, 3rd edn. Blackie, London.

Fowles, G. R. (1975) *Introduction to modern optics*, 2nd edn. Holt, Rinehart and Winston, New York.

Françon, M. (1966) *Diffraction: coherence in optics*, English edn. (transl. Jeffrey, B. and edited Sanders, J. H.). Pergamon, London.

Ghatak, A. (1977) *Optics*. Tata McGraw-Hill, New Delhi.

Goodman, J. W. (1968) *Introduction to Fourier optics*. McGraw-Hill, New York.

Hecht, E. and Zajac, A. (1974) *Optics*. Addison-Wesley, Reading, Mass.

Jenkins, F. A. and White, H. E. (1976) *Fundamentals of optics*. 4th edn. McGraw-Hill Kogakusha, Tokyo.

Klein, M. W. (1970) *Optics*. Wiley, New York.

Lipson, S. G. and Lipson, H. (1981) *Optical physics*, 2nd edn. CUP, London.

Longhurst, R. S. (1973) *Geometrical and physical optics*. 3rd edn. Longman, London.

Nussbaum, A. and Phillips, R. A. (1976) Contemporary optics for scientists and engineers. Prentice-Hall, New Jersey.

Smith, F. G. and Thomson, J. H. (1971) *Optics*. Wiley, London.

Stone, J. M. (1963) *Radiation and optics*. McGraw-Hill, New York.

Taylor, C. A. (1978) *Images*. Wykeham Publications, London.

Welford, W. T. (1981) *Optics*, 2nd edn. OUP, Oxford.

CHAPTERS 1 AND 2

Hecht, E. and Zajac, A. (1974) *Optics*. Addison-Wesley, Reading, Mass.

Jenkins, F. A. and White, H. E. (1976) *Fundamentals of optics*. 4th edn. McGraw-Hill Kogakusha, Tokyo.

Longhurst, R. S. (1973) *Geometrical and physical optics.* 3rd edn. Longman, London.

Taylor, C. A. (1978) *Images.* Wykeham Publications, London.

CHAPTERS 3 AND 4

Bracewell, R. N. (1979) *The Fourier transform and its applications,* 2nd edn. McGraw-Hill, New York.

Champney, D. C. (1985) *Fourier transforms in physics.* Adam Hilger, Bristol and Boston.

Gaskill, J. D. (1978) *Linear systems, Fourier transforms and optics.* Wiley, New York.

Wiener, N. (1933) *The Fourier integral and certain of its applications.* CUP, London.

CHAPTER 5

Barnes, K. R. (1971) *The optical transfer function.* Hilger, London.

Bragg, W. L. (Phillips, D. C. and Lipson, H. (eds.)) (1975) *The development of X-ray analysis.* Bell, London.

Casasent, D. (ed.) (1978) *Optical data processing.* Springer, Berlin.

Culling, C. F. A. (1974) *Modern microscopy: elementary theory and practice.* Butterworth, London.

Cutrona, L. J., Leith, E. N., Palermo, C. J., and Porcello, L. J. (1960) Optical data processing and filtering systems. *IRE Trans. Inform. Theory,* **IT-6**, 386–400.

Duffieux, P. M. (1946) *L'intégrale de Fourier et ses applications à l'optique.* Published privately: Rennes, France.

Duffieux, P. M. (1983) *The Fourier transform and its applications to optics,* 2nd edn. Wiley, New York.

Françon, M. (1974) *Holography,* English edn. (transl. Spruch, G. M.). Academic Press, New York.

Françon, M. (1979) *Optical image formation and processing,* English edn. (transl. Jaffe, B. M.). Academic Press, New York.

Gabor, D. (1972) (*Edwards Memorial Lecture*) *Holography: 1948–1971.* The City University, London.

Graham, N. (1979) Does the brain perform a Fourier analysis of the visual scene? *TINS,* **2**, 207–208.

Harburn, G., Taylor, C. A. and Welberry, T. R. (1975) *Atlas of optical transforms.* Bell, London.

Hawkes, P. W. (ed.) (1980) *Computer processing of electron microscope images.* Springer, Berlin.

Hopkins, H. H. (1962) 21st Thomas Young Oration. The application of frequency response techniques in optics. *Proc. Phys. Soc.,* **79**, 889–919.

Kihlborg, L. (ed.) (1979) *Nobel Symposium 47. Direct imaging of atoms in crystals and molecules.* The Royal Swedish Academy of Sciences, Stockholm.

Lee, S. H. (ed.) (1981) *Optical information processing — fundamentals.* Springer, Berlin.

Lipson, H. S. (1970) *Crystals and X-rays.* Wykeham Publications, London.

Lipson, H. (ed.) (1972) *Optical transforms.* Academic Press, London.

Lipson, H. and Taylor, C. A. (1958) *Fourier transforms and X-ray diffraction.* Bell, London.

Marathay, A. S. (1982) *Elements of optical coherence theory.* Wiley, New York.

McLachlan, D. (1962) The role of optics in applying correlation functions to pattern recognition. *J. Opt. Soc. Amer.*, **52**, 454–459.

Mertz, L. (1965) *Transformations in optics.* Wiley, New York.

Michelson, A. A. (1927) *Studies in optics.* University of Chicago Press, Chicago. (Phoenix Science Series edn, 1962.)

O'Neill, E. L. (1956) Spatial filtering in optics. *IRE Trans. Inform. Theory*, **IT-2**, 56–65.

O'Neill, E. L. (1963) *Introduction to statistical optics.* Addison-Wesley, Reading. Mass.

Rogers, G. L. (1954) The Abbe theory of microscopic vision and the Gibbs phenomenon. *Amer. Jnl. Phys.*, **22**, 384–389.

Schawlow, A. L. (ed.) (1969) *Lasers and light* (Readings from *Scientific American*). Freeman, San Francisco.

Shannon, C. (1949) *The mathematical theory of communication.* University of Illinois, Urbana.

Shulman, A. R. (1970) *Optical data processing.* Wiley, New York.

Smith, H. M. (1975) *Principles of holography*, 2nd edn. Wiley, New York.

Solem, J. C. and Baldwin, G. C. (1982) Microholography of living organisms. *Science*, **218**, 229–235.

Stark, H. (ed.) (1982) *Applications of optical Fourier transforms.* Academic Press, New York.

Stokes, A. R. (1952) Three-dimensional diffraction theory of microscope image formation. *Proc. Roy. Soc. A*, **212**, 264–274.

Stroke, G. W. (1969) *An introduction to coherent optics and holography*, 2nd ed. Academic Press, New York.

Taylor, C. A. and Lipson, H. (1964) *Optical transforms.* Bell, London.

Wiener, N. (1949) *The extrapolation, interpolation and smoothing of stationary time series.* Technology Press and Wiley, New York.

CHAPTER 6

Andrew, E. R. (1985) Nuclear magnetic resonance imaging in medicine: physical principles. *Proc. Roy. Soc. Lond.*, **B225**, 399–410.

Bergland, G. D. (1969) A guided tour of the Fast Fourier Transform. **IEEE Spectrum**, **6**, 41–52.

Cannon, T. M. and Hunt, B. R. (1981) Image processing by computer. *Sci. Amer.*, **245**, 136–145.

Foster, M. A. and Hutchison, J. M. S. (1985) NMR imaging — method and applications. *J. Biomed. Eng.*, **7**, 171–182.

Gordon, R., Herman, G. T., and Johnson, S. A. (1975) Image reconstruction from projections. *Sci. Amer.*, **233**, 56–68.

Green, W. (1984) *Digital image processing: a systems approach*. Van Nostrand.

Hendee, W. R. (1983) *The Physical principles of computed tomography*. Little, Brown and Co., Boston.

Herman, G. T. (ed.) (1979) *Image reconstruction from projections*. Springer, Berlin.

Herman, G. T. (1980) *Image reconstruction from projections: the fundamentals of computerised tomography*. Academic Press, New York.

Morgan, C. L. (1983) *Basic principles of computed tomography*. University Park Press, Baltimore.

Raviv, J., Greenleaf, J. F., and Herman, G. T. (eds). (1979) *Computer aided tomography and ultrasonics in medicine*. North-Holland, Amsterdam.

Wells, P. N. T. (1984) Medical imaging *Phys. Bull.*, **35**, 193–195.

Whittingham, T. A. (1984) Ultrasound in medical diagnosis. *Phys. Bull.*, **35**, 196–198.

CHAPTER 7

Bahcall, J. N. and Spitzer, L. (1982) The space telescope. *Sci. Amer.*, **247**, 38–49.

Bell, R. J. (1972) *Introduction to Fourier transform spectroscopy*. Academic Press, New York.

Chown, M. (1985) A radio telescope bigger than the Earth. *New Scientist*, **105**, 29–32.

Françon, M. (1966) *Optical interferometry*, English edn. (transl. Wilmanns, I.). Academic Press, New York.

Griffiths, P. R. (ed.) (1978) *Transform techniques in chemistry*. Heyden, London.

Hanbury Brown, R. (1974) *The intensity interferometer*. Taylor and Francis, London.

Hewish, A. (1965) The synthesis of giant radio telescopes. *Sci. Progr.*, **53**, 355–368.

Henbest, N. (1982) Jodrell under Merlin's spell. *New Scientist*, **96**, 442–448.

Hey, J. S. (1971) *The radio universe*. Pergamon, Oxford.

Hey, J. S. (1973) *The evolution of radio astronomy*. Elek Science, London.

Jennison, R. C. (1966) *Introduction to radio astronomy*. Newnes, London.

Readhead, A. C. S. (1982) Radio astronomy by very-long-baseline interferometry. *Sci. Amer.*, **247**, 39–47.

Smith, F. G. (1966) *Radio astronomy*, 3rd edn. Penguin Books, London.

Solutions to problems

CHAPTER 1

1.1 Eqn (1.04) $D = 0.5\,\text{mm}$

1.2 Eqn (1.06) $D < 6.3 \times 10^{-2}\,\text{mm}$

1.3 Eqn (1.06) $\phi < 550 \times 10^{-6}\,\text{rads}$

1.4 Derivation like that for Eqn (1.12). $\lambda/2$ in (1.13) not needed because there is no reflection at an air/glass interface involved.

CHAPTER 2

2.1 3rd zero when $a \sin \theta = 3\lambda$ (§2.2)
  $x = 100 \tan \theta = 0.16\,\text{cm}$

2.2 Eqn (1.06) $\phi < \lambda/D$, $D = 1\,\text{mm}$
  Source width $W = f\phi$. $W < 0.01\,\text{cm}$

2.3 The light patch, diameter x say on the Moon, can be taken as the Airy disc of a diffraction limited beam of initial diameter $2\,\text{mm}$.
  $\theta_{\text{rads}} \simeq 1.22\lambda/a$ Eqn (2.07)

  $x = 2r\theta_{\text{rads}}$, $r = 376 \times 10^3\,\text{km}$

  $x \simeq 290\,\text{km}$.

2.4 Similar to 2.3. Area $= 0.018_7\,\text{km}^2$.

2.5 See §2.5. Grating max. of order n, when $D \sin \phi = n\lambda$.
  Individual slit zero intensity when $a \sin \theta = m\lambda$.
  Missing orders when these coincide, i.e. when $n = mD/a$.
  Ans. $n = 5, 10, 15, \ldots$
  Intensity of diffraction max.

$$\propto \left(\frac{\sin \frac{\pi n a}{D}}{\frac{\pi n a}{D}} \right)^2 \quad \text{Eqn. (2.14)}$$

$$I_{n=0} \propto 1$$

$$I_{n=3} \propto \left(\frac{\sin\frac{3\pi}{5}}{\frac{3\pi}{5}}\right)^2 = 0.25_4$$

$$I_{n=3}/I_{n=0} = 0.25_4 .$$

2.6 Total amplitude, from one repeat, in the direction of a diffraction maximum, is

$$= \int_D A \sin\frac{2\pi m x}{D}\, e^{i(2\pi/\lambda)x\sin\theta}\,dx \qquad \text{(cf. text related to Fig. 4.03)}$$

$$= 0 \quad \text{unless} \quad \frac{m}{D} = \frac{\sin\theta}{\lambda} . \text{ (Put sine into exponential notation.)}$$

$D\sin\theta = n\lambda$ gives the direction of diffraction order n (Eqn (2.10)). Therefore only the pair of diffraction maxima of order $n = m$ is given by this grating.

2.7 Phasor diagram is a closed hexagon. $2\beta = 2\pi/6$ (Fig. 2.10(b)).
$2\beta = (2\pi/\lambda)D\sin\theta.$ $D = 6\lambda,$

$$\therefore \sin\theta = 1/36, \quad \theta = 1.59_2^\circ$$

Diffraction max. when $D\sin\theta = n\lambda$
$n = \pm 1 \quad \sin\theta = 1/6 \quad \theta = \pm 9.6^\circ$

$$\text{Intensity} \propto \left(\frac{\sin\frac{\pi}{6}}{\frac{\pi}{6}}\right)^2 = 0.91$$

$n = \pm 2 \quad \sin\theta = 1/3 \quad \theta = \pm 19.5^0$

$$\text{Intensity} \propto \left(\frac{\sin\frac{\pi}{3}}{\frac{\pi}{3}}\right)^2 = 0.68$$

CHAPTER 3

3.1 Graph resembles Fig. 3.01(a).
Procedure follows Fig. 3.02. $A_0 = 6$, $a_{n=2} = -1$, $a_{n=3} = 0$.
All $b_n = 0$ (there are no sine components in $f(x)$).

3.2

$$f(x) = \sin\frac{2\pi 5x}{D} + \frac{i}{2}\cos\frac{2\pi 5x}{D}$$

$$= \frac{1}{2i}\left(e^{i\frac{2\pi 5x}{D}} - e^{-i\frac{2\pi 5x}{D}}\right) + \frac{i}{2}\frac{1}{2}\left(e^{i\frac{2\pi 5x}{D}} + e^{-i\frac{2\pi 5x}{D}}\right)$$

$$= e^{i\frac{2\pi 5x}{D}}\left(\frac{1}{2i} + \frac{i}{4}\right) + e^{-i\frac{2\pi 5x}{D}}\left(\frac{3i}{4}\right)$$

$$= e^{i\frac{2\pi 5x}{D}}\left(-\frac{i}{4}\right) + e^{-i\frac{2\pi 5x}{D}}\left(\frac{3i}{4}\right) \tag{1}$$

(a)   The above shows that the only terms in the exponential Fourier series
(Eqn (3.10)) for $f(x)$ are those for $n = \pm 5$, with $C_5 = i/4$ and $C_{-5} = 3i/4$.
(b)   Substitution of (1) for $f(x)$ in Eqn (3.13) shows that the only non-zero
C_n are for $n = \pm 5$, and that their values are as above.

CHAPTER 4

4.1 The apertures are now at $x = 0$ and $x = D$. Relative to Fig. 2.08(b) the
phasor diagram is rotated through $\beta = \pi u D$ (Eqn 2.11).

$$\delta\text{-functions at } x = 0 \text{ and } x = D \text{ give}$$

$$F(u) = 2\,e^{2\pi iuD/2}\cos\pi uD \qquad (\S 4.3)$$

where the exponential contains a phase of the same amount.

4.2 (a)   A sketch of nine dots 'weighted' towards the centre of the row.

(b)   A sketch of the following pattern

where the sizes of the spots indicate their 'weighting'.

(c)   A square 5×5 matrix of dots.

4.3 $F(u) = e^{-2\pi iuD} + e^0 + e^{+2\pi iuD}$

$= 1 + 2\cos 2\pi uD.$

Intensity is the square of this and it has to be multiplied by the single-aperture intensity term. (See Fig. 2.12(c).)

4.4 Following the same procedure as in (a) → (c) in Fig. 4.05, an ordinate x_2 in the input, i.e. $f(x_2)$, when smeared, becomes the function $f(x_2)g(x - x_2)$ in the output. The contribution this makes (i.e. its ordinate) at x_1 in the output is therefore $f(x_2)g(x_1 - x_2)$. This is the same as the ordinate in the shaded area in (e) at x_2, viz. $f(x_2)g(x_1 - x_2)$. This perhaps seemingly strange property of the shaded area in (e) is due to $g(x)$ being 'folded' (reversed) in the mathematical derivation. The value of $g(x_1 - x)$ at x_2 is the same as the value of $g(x - x_2)$ at x_1.

4.5 (a)   In the convolution, either function can be reversed. Often the choice influences the ease of solution — a matter for experience and trial. It is probably best here to reverse the signal and evaluate

$$r(\tau) = \int_\alpha^\beta \frac{1}{\Delta} e^{-t/\Delta} \cdot \frac{(\tau - t)}{\Delta} \cdot dt$$

where α, β are appropriate limits. Two sets of limits have to be taken, for the reversed signal starting before or after $t = 0$. In the first case $\alpha = 0$, $\beta = \tau$; in the second $\alpha = \tau - \Delta$, $\beta = \tau$.

The integration is straightforward — use integration by parts

(i) $r(\tau) = \dfrac{\tau}{\Delta} - \{1 - e^{-\tau/\Delta}\}$, $0 < \tau \leq \Delta$

$\tau > \Delta$

(ii) $r(\tau) = e^{-\tau/\Delta}$

The solution is shown below:

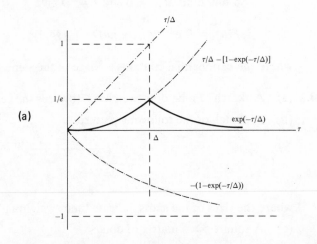

(b)   Ten strips 0.1 second wide gives surprisingly accurate results. Each strip sets up an impulse response weighted by its area — see diagram. To find the response at any time draw a vertical line at that value of time, and add the ordinates of all the impulse responses crossed. As an example, at $t = 0.55$ second, the ordinates to be added are

$$0.045\,e^{-0.05} + 0.035\,e^{-0.15} + 0.025\,e^{-0.25} + 0.015\,e^{-0.35} + 0.005\,e^{-0.45} \ .$$

A pocket calculator then gives $r(0.55) = 0.106$. The algebraic solution is

$$r(0.55) = 0.55 - 1 + e^{-0.55} = 0.127$$

(b)

CHAPTER 7

7.1 §7.3 θ very small.

$$2(h_2 - h_1) = k\lambda_1 = (k+1)\lambda_2$$

$$2(h_2 - h_1) = \frac{\lambda_2}{\lambda_1 - \lambda_2}\lambda_1$$

$$h_2 - h_1 \quad \simeq 0.03\,\text{cm}.$$

7.2 With $B(\phi)$ constant, the expression for $V(D)$ becomes

$$V(D) = \frac{\displaystyle\int_{-\phi_0/2}^{+\phi_0/2} \cos(2\pi D\phi/\lambda)\,d\phi}{\displaystyle\int_{-\phi_0/2}^{+\phi_0/2} d\phi}$$

$$= \frac{\lambda/2\pi D \left[\sin(2\pi D\phi/\lambda)\right]_{-\phi_0/2}^{+\phi_0/2}}{\left[\phi\right]_{-\phi_0/2}^{+\phi_0/2}}$$

$$= (\lambda/2\pi D)\frac{2\sin\dfrac{\pi\phi_0 D}{\lambda}}{\phi_0}$$

$$V(D) = \frac{\sin\dfrac{\pi\phi_0 D}{\lambda}}{\dfrac{\pi\phi_0 D}{\lambda}} \ .$$

Index